Doctrine in Shades of Green

Doctrine in Shades of Green

Theological Perspective for Environmental Ethics

ANDREW J. SPENCER

WIPF & STOCK · Eugene, Oregon

DOCTRINE IN SHADES OF GREEN
Theological Perspective for Environmental Ethics

Copyright © 2022 Andrew J. Spencer. All rights reserved. Except for brief quotations in critical publications or reviews, no part of this book may be reproduced in any manner without prior written permission from the publisher. Write: Permissions, Wipf and Stock Publishers, 199 W. 8th Ave., Suite 3, Eugene, OR 97401.

Wipf & Stock
An Imprint of Wipf and Stock Publishers
199 W. 8th Ave., Suite 3
Eugene, OR 97401

www.wipfandstock.com

PAPERBACK ISBN: 978-1-6667-0225-5
HARDCOVER ISBN: 978-1-6667-0226-2
EBOOK ISBN: 978-1-6667-0227-9

MARCH 8, 2022

Unless otherwise marked, Scripture quotations are from the ESV® Bible (The Holy Bible, English Standard Version®), copyright © 2001 by Crossway, a publishing ministry of Good News Publishers. Used by permission. All rights reserved.

Contents

Preface | vii

Acknowledgments | xi

PART ONE   THE GROUNDWORK

CHAPTER ONE   Introduction | 3
CHAPTER TWO   Four Doctrinal Questions | 17

PART TWO   ECOTHEOLOGY

CHAPTER THREE   An Ecotheological Perspective for Environmental Ethics | 41
CHAPTER FOUR   The Ecotheology of Ernst Conradie | 65

PART THREE   A LIBERAL ENVIRONMENTAL ETHICS

CHAPTER FIVE   A Liberal Perspective for Environmental Ethics | 89
CHAPTER SIX   The Liberal Environmentalism of Joseph Sittler | 119

PART FOUR   A FUNDAMENTALIST ENVIRONMENTAL ETHICS

CHAPTER SEVEN   A Fundamentalist Perspective for Environmental Ethics | 139

PART FIVE   AN EVANGELICAL ENVIRONMENTALISM

CHAPTER EIGHT   An Evangelical Perspective for
Environmental Ethics | 181

CHAPTER NINE   The Environmentalism of Francis Schaeffer | 209

Conclusion | 223

Bibliography | 227

Index | 245

Preface

I DIDN'T CHOOSE ENVIRONMENTAL ETHICS; environmental ethics chose me. This is at least partially true. My interest in questions of the environment stems from growing up too close to the cleanup site from a nuclear fuel rod reprocessing plant (In high school we used to cheer, "Heck no, we don't glow," in response to jibes from other teams) and a still-questionable decision get on a bus one chilly morning in the autumn of 2000. The rest, as they say, is history.

The nuclear site was in West Valley, New York, which is about as rural as you can get without heading to Big Sky country. The nuclear fuel rod reprocessing plant had been shut down by Jimmy Carter's SALT II treaty, which left the local economy somewhere between depressed and desperate. I got sucked into the local resistance movement against putting a low-level radioactive waste storage facility on the land that was now basically vacant and pretty much clean, except where a certain amount of nuclear waste was awaiting storage for shipment to permanent storage at Yucca Mountain, which was still a hope on the horizon. Westinghouse had the contract to run the facility—the West Valley Demonstration Project—that led to decent funding for STEM curriculum in a town that could have served as a backdrop to *October Sky* except without the coal mine to provide even a decent hope of a living.

I made it to the United States Naval Academy with every intention of becoming an engineer, but I got lured into the humanities and developed a love for literature. However, even my degree in English led to a Bachelor of Science, so I found myself pacing outside a bus one day, knowing that if I got on it and if I passed my interviews at Naval Reactors, I would have to become an officer on a nuclear submarine. About a year later I found myself in Nuclear Power School learning the deep mysteries of what makes a reactor work. Later, after I left the Navy, I took a job as an instructor at a commercial nuclear power plant, teaching operators how to operate a large, commercial plant. It was not a bad way to work your way through seminary.

Environmental ethics chose me because I had developed an abiding interest in technology, an insider's understanding about electricity generation, and was immersed in a religious tradition that had powerful tools to engage in environmental ethics but seemed to have political reasons not to. Thankfully, my seminary ran a special colloquium on the environment, which further developed my interest. That same seminary also included a recurring elective on environmental ethics in the MDiv curriculum, which was where I first began to see the wide range of positions on creation care among professing Christians and to explore the theological ideas that drive ethicists toward sometimes radically different ethical proposals.

Critics tend to blame politics for the division among Christians, particularly among the theologically conservative US Christians with whom I am most familiar. I have no doubt that certain issues have been clustered together in an unhelpful manner. Thus, support for abortion is often associated with environmentalism, and advocacy for nuclear power is typically associated with corporate cronyism. It is tempting to lump these issues into the Left-Right divide in US politics and walk away, but when the scope of research expands beyond the borders of the US, there are still radical differences in the way environmentalists approach various issues. After years of reading religious perspectives on environmental ethics, I began to see certain patterns of theology that coincided with specific approaches to the environment.

My dissertation topic and thesis crystalized after I read an article by David Horrell, Cherryl Hunt, and Christopher Southgate that analyzed the use of Scripture in environmental ethics across three basic theological streams.[1] The sense that my own position had not been precisely analyzed led me to think more on the approach, as I sought to distance my own conservative evangelical hermeneutic from that of Hal Lindsey's approach in *The Late Great Planet Earth* which led me to think through the range of literature I had and was reading. The article is largely correct about the typology of hermeneutics, each type relying upon a particular understanding of revelation, but there are also divisions among theological perspectives in several other key doctrines. As I continued to explore

---

1. Horrell, Hunt, and Southgate, "Appeals to the Bible," 219–38.

the literature of the field, I noticed a recurring pattern of four doctrines that mark points of difference between distinct types of Christian environmentalism. Thus, a thesis was born.

Acknowledgments

Many thanks to my wife, who did not know I was working on converting my dissertation into a book, but still put up with me working in my basement office when I should have been writing my next book. Thank you for your patience with my writing. Thank you, as well, for reading the earlier drafts of this project, which were submitted as my dissertation.

Grace, Freeman, and Martha also deserve thanks for bearing with their father working a full-time job and writing a lot more than he should. May you someday pick this book up and be helped by it. Maybe you will even find it helpful as you sort through your faith in light of the ethical demands of the day. Keep to the old paths.

My thanks go out to Daniel Heimbach, who helped shepherd my dissertation through various stages and whose careful approach to ethics has helped shape my own. John Frame and Mark Liederbach served as members of my dissertation committee and affirmed the project, which bolstered my confidence to move this from the shadows of ProQuest into the light of publication. At the end of the day, of course, any errors and failings that remain in this project are my own.

I am grateful for several journal editors who granted me permission to reuse portions of published articles again. It is good not to have to rewrite pieces that built the dissertation in the first place.

After all is said and done, I am grateful to you, reader, who have picked this book up from among many other options. I hope you find it helpful.

Part One

THE GROUNDWORK

CHAPTER ONE

INTRODUCTION

IN THE FLOOD OF green messages that surround us in Western culture every day there are basic assumptions that people make as they classify each other. "Ecofriendly" means politically and theologically liberal. "Conservative" means denying climate change, driving a big SUV (but let's be honest, a lot of liberals drive these, too), and trying to own the libs by not recycling. That is, at least, the messaging that we get bombarded with.

The problem with consistent messaging is that eventually people start to believe it, even if it requires denying the logical results of one's core beliefs. The stereotype sometimes becomes a reality and things that are "off normal" strike us as odd. As a result, people are sometimes surprised to find out that I go to the small Baptist church down the road that holds to the inerrancy of Scripture, but I also have solar panels on my roof and pollinator gardens in my yard. Just to confirm that I'm weird, I usually let people know that I keep my house at about the recommended 63°F in the winter and 78°F in the summer. My theological convictions drive my environmental ethics and lead me to invest time, money, and energy into choices that tend to be much more popular on the other side of the political and religious divide.

This book wrestles with the theological underpinnings of environmental ethics from four distinct theological perspectives along the spectrum of Christian belief. One goal of this book is to show that a wide range of Christian theology can support an earth-positive environmental ethics. Therefore, there is little need for people to change their doctrines to suit the demands of ecology. At the same time, though many people from different forms of Christianity may arrive at similar ecological practices, we should not gloss over the real theological differences, which explain some of the divergences in policy preferences between, say, a theological liberal and an evangelical. A second goal of this book is to provide some points of contact between theological streams that can lead to better cooperation where possible and better dialogue where differences remain.

THESIS

In this book, I seek to show that divergences between Christian environmental ethics are largely explained by responses to four particular theological questions. Together, these four questions form a theological perspective for environmental ethics. The contents of a theological perspective provide a structure for analysis, which can simplify complexity and illuminate obscurity in the ongoing debate over the appropriate Christian approach to environmental ethics. This book examines the theological perspective for environmental ethics of four different theological streams that diverge along a theological spectrum. The object is to provide a common framework for engagement between theological streams rather than to critique any particular perspective.

There has already been enough blame-throwing to last a few millennia. For example, division among Christians on the issue of the environment has been exacerbated by Lynn White's historic essay that blames a Christian worldview for ecological degradation. Since "The Historical Roots of Our Ecological Crisis" was published in 1967, the Christian environmental debate has been framed as a response to White's thesis.[1] Some scholars are coming to recognize how stultifying that approach has become to meaningful dialogue on environmentalism.[2] Among orthodox Christians, a great deal of ink has been spilt in defending Christianity against the attack levied by White. This defense has undermined positive efforts on environmental ethics, so that positive theological statements have lagged behind.[3] The constant battling over whether Christianity is good for the environment means that little attention has been paid to the specific theological methods used to develop Christian environmental ethics and few attempts have been made to understand the theological

---

1. White, "Historical Roots," 15–31.
2. Danielsen, "Fracturing over Creation Care?," 202.
3. Fowler, *Greening of Protestant Thought*, 19–20.

structures in place beneath those environmental ethics. Lately, battles over the so-called Green New Deal have made it even more difficult to begin talking about environmental ethics because of emotionally charged divisions within the culture.

A survey of theological perspectives of Christian environmentalists is an important pursuit at this point in Christian history because of the growing distance between the ethical formulations, driven by diverse theological methods, of various streams of Christianity. Paul Allen notes this divergence of theological method, arguing,

> Twentieth-century Christian theology has moved decisively yet sporadically in a number of contradictory directions. The sporadic character of theology is most deeply felt in terms of theological method. As theological scholarship widened to include a global scope of cultures and traditions to which the academy and churches were previously indifferent, theology took on a vastly more pluralistic tone.[4]

According to Allen, "Twentieth-century theology has, by and large, adapted the theological developments of the nineteenth century to a culture that is no longer habitually Christian."[5] At this point, early in the twenty-first century, these sentiments indicate a need for an examination of the theological methods that support various environmental ethics.

An examination of theological methods used by diverse environmental ethicists, with due consideration of theological foundations, will enrich the understanding of environmental ethics among Christians. But full-scale analyses for each interlocutor in such a widespread debate will not be done here for two reasons. First, because many of those who write about environmental ethics from a Christian viewpoint have not written sufficiently to substantiate thorough analysis of their theological methods. Second, even if there were sufficient material to analyze, a complete assessment of each writer's theological method goes beyond what this project requires. Instead of complete analysis, theological perspective offers a tool for an abbreviated theological assessment, tailored to the subject matter, which can form a structure for diagnosis and dialogue.

---

4. Allen, *Theological Method*, 167–68.
5. Allen, *Theological Method*, 168.

THEOLOGICAL PERSPECTIVE

I use the term "theological perspective" to focus on the specialized questions of contact between a theological foundation and a particular issue. As with any specialized vocabulary, there are difficulties associated with selecting a phrase for a special purpose. The first difficulty is semantic overlap with existing theological terminology. The second difficulty is providing a sufficiently clear definition of the term itself. Theologian Richard Lints provides a helpful approach to theological method, which I have here adapted to the concept I call "theological perspective."

In his book *Fabric of Theology*, Lints develops the concepts of "theological vision" and "theological matrices," upon which my term—"theological perspective"—is built. According to Lints, each individual has a unique worldview, which invokes a specific "theological framework."[6] He argues,

> We might picture a person's perspective on his or her experiences as something like a map or matrix on which certain experiences and actions are placed or arranged. The significance of any given experience would in part be a function of where it was placed on the map. And the shape and experience or action would be placed on it. An individual hearing certain words or seeing certain objects would grant them a significance determined on the basis of the role they played on the map, on the basis of the relationship to other words or experiences. These new experiences might "fit together" or they might be "in conflict," but in any event it would be the matrix that would inform the individual about what it means for experiences and action to "fit together" and to be in "conflict."[7]

These matrices function as a subset of an overall theological vision, with a different matrix coming into play depending on the experience under consideration.[8]

According to Lints, "The matrix is neither identical with nor separate from the experiences and actions. Yet the matrix is absolutely essential for understanding the experiences and actions."[9] However, "The meaning of any given experience is not entirely determined by the matrix

---

6. Lints, *Fabric of Theology*, 12–13.
7. Lints, *Fabric of Theology*, 13.
8. Lints, *Fabric of Theology*, 15.
9. Lints, *Fabric of Theology*, 13.

even though it is greatly influenced by it. Likewise, the matrix both influences and is influenced by the experiences."[10] There is in Lints's accounting a two-way shaping going on. A matrix is a cognitive element in a worldview; thus it can be intentionally reshaped by those who recognize the existence of it. However, people who "remain largely ignorant of their matrices will be the group most likely to be controlled by them."[11] Thus, there seems to be warrant to identify theological matrices so they can be evaluated, modified, replaced, or discarded.

A theological vision is constructed of a matrix of matrices.[12] Lints's purpose in *The Fabric of Theology* is to examine the theistic matrix, which is comprised of a specific set of beliefs about God and his nature. So, Lints explains, "The belief that God exists, for example, is normally accompanied by a host of other beliefs—what that God is like, whether that God can and does communicate, how that God is known, and what that God requires of us."[13] In *The Fabric of Theology*, Lints tackles beliefs that are central to Christianity, explaining the basis for a particularly evangelical theology. The concept Lints presents is useful beyond the study of theology proper, and his basic formulation can be applied to issues like environmental ethics. This project uses Lints's basic model of a theological matrix as a foundation for the concept of a theological perspective.

As I use it, "theological perspective" is not a generic synonym for "worldview" or a sweeping term for "thinking theologically." Worldview is inherently theological, and it is important for environmental ethics. Indeed, a presupposition of this project is that, whether conscious of it or not, all people filter the world around them through a particular theological lens.[14] A less controversial assertion, since this project focuses on four Christian theological perspectives, is that Christians engage the world in an inherently theological manner. The way that a Christian perceives the world and responds tends to be consistent with her actual theological beliefs, even if not the stated theological beliefs. To analyze environmental ethics through worldview would be too broad an approach to be helpful. Instead, theological perspective narrows the focus to four particular

---

10. Lints, *Fabric of Theology*, 14.
11. Lints, *Fabric of Theology*, 14.
12. Lints, *Fabric of Theology*, 19.
13. Lints, *Fabric of Theology*, 18.
14. Smith, *Fall of Interpretation*, 159–98.

Prismatic Analogy

If life, which is undergirded by a theological vision, is represented as a complex scientific exploration to discover the properties of light, then the interaction between theological perspective and a particular ethical issue can be related to the physical analogy of a prism. The purpose of this theological-optical experiment, then, is to achieve a discernable spectral output where effective analysis of both the ethical subject (e.g., environmental ethics) and theological interpretation of that topic can be achieved. The material composition of the prism is like the theological perspective, as it is defined for this project. The incoming light into the prism is analogous to the ethical subject under consideration.

The material composition of a prism impacts its index of refraction. The index of refraction is specific to the material and is a mathematical factor by which the speed of light is reduced as it passes through a physical medium in comparison to the absolute speed of light, which would be evidenced only in a perfect vacuum.[15] In this analogy, theological perspective is like the material characteristics of the prism that impact the index of refraction. Different theological perspectives have particular constituent doctrinal components, which affect the manner in which ethical subjects are handled.

The amount of refraction of light by a prism is also dependent upon the frequency of the incoming light; higher-frequency light waves (toward the blue end of the spectrum) refract more, while lower-frequency light waves (toward the red end of the spectrum) refract less. This is called dispersion and explains why white light can turn into a rainbow when passing through a prism.[16] The ethical subject under consideration is analogous to the frequency composition of the incoming light. Ethical subjects are rarely (if ever) monochromatic (i.e., having only one frequency of light) and typically involve multiple concerns; as the output of the prism depends on the nature of the light that shines on it, so is the

---

15. Band, *Light and Matter*, 100.
16. Band, *Light and Matter*, 107.

outcome of an ethical analysis dependent on the nature and complexity of the ethical issues presented.[17]

Divergences between Christian environmental ethics are rooted in the response to four key theological questions. These questions form a theological perspective. This project seeks to demonstrate how four theological questions have particular bearing on environmental ethics by showing how, like a common input of light in a controlled environment, examining the responses of theologians to the four questions that make up a theological perspective for environmental ethics has special analytic value for discussions of environmental ethics. The point, then, is to demonstrate the value of intentionally assessing the theological perspective when evaluating ethical argumentation in public discourse, so that dialogue can occur that is related to the actual theological differences (i.e., material composition of the prism) instead of variations in other conditions.

In other words, this book is like a scientific experiment where I am attempting to isolate one factor (i.e., theological perspective represented by material composition of the prism) to demonstrate the value in beginning discussions in environmental ethics by considering theological perspective rather than talking past others because other factors vary widely outside of an experimental environment. In some ways I am attempting to do what Sir Isaac Newton did for the study of optics and light. He writes, "My Design in this Book is not to explain the Properties of Light by Hypotheses, but to propose and prove them by Reason and Experiments."[18] One man's reason and experiments is another man's theological research in book format.

A TYPOLOGY OF THEOLOGICAL PERSPECTIVES

This project examines four typological categories of Christian theology as case studies in the larger experiment of establishing the elements of a theological perspective for environmental ethics. The types themselves

---

17. There are more factors that go into the relationship, including impurities in the prism (theological inconsistencies), the temperature of the glass (cultural climate), angle of incidence (manner in which the information is presented), etc. However, my analogy is sufficient to demonstrate the similarities and help explain the thesis without overly complicating the imagery. For a concise explanation of the properties of optical prisms and their relationships, see Chartier, *Introduction to Optics*, 131–35.

18. Newton, *Opticks*, 1.

are very different streams of Christian theology. To return to the prismatic analogy, by maintaining other elements of the experiment consistent, the thesis that there is a theological perspective for environmental ethics can be tested. With heuristic intent, the four typological categories of Christian environmental ethics that will be evaluated in this project are the ecotheological, liberal, fundamentalist, and evangelical theological perspectives for environment ethics. Although such categorical labels are, at times, used as epithets in theological banter, in this project the intention is to accurately describe the categories and use them for the sake of simplicity of expression. Each of the types will be defined in greater detail in the chapter dealing with that theological category, but a brief outline is in order here.

Ecotheology is a version of liberation theology that defines the oppression of the earth as the ultimate concern for Christian theology. Like other streams of liberation theology, ecotheology is characterized by a manner of interpreting the text of Scripture and the traditional formulations of Christianity in a way that is inconsistent with most orthodox readings. Ecotheology largely focuses on praxis, which is the practical application of a Christian ethic to ecological issues, at the expense of a focus on central truths of the gospel. In fact, in ecotheology the gospel becomes the outworking of an environmental ethics rather than a timeless truth that is essential for salvation. This view is more often represented among very progressive Christians, though recent efforts have been made to present it as a viable option for evangelicals.

Liberal theologians tend to hold environmental activism as a major component of Christianity, though not essential to salvation. According to this view, Scripture can be interpreted in an environmentally friendly manner. While there are elements of traditional Christian theology that require revision to suit contemporary environmental concerns, Christianity is a useful motivator for environmentalism. Creation has value largely independent of external considerations. Scripture teaches that humans are to serve the created order for the glory of God, and to interfere with the environment minimally. A central duty of Christians in the liberal theological perspective for environmental ethics is to preserve creation in as pure a state as possible.

There are few, if any, fundamentalist Christians who have positively engaged the topic of environmental ethics. However, using basic theological texts from the fundamentalist tradition, a theological perspective for environmental ethics from a fundamentalist understanding can be developed. As it stands, most fundamentalist statements on environmentalism describe human concern for the environment as a matter of personal choice, but one that is mainly founded on utility. A theological formulation of Christian environmentalism is considered a distraction from the greater purpose of evangelizing the masses. Working to improve the environment might not be bad. However, as one professor once wrote on one of my papers, "It's like shuffling deck chairs on the Titanic." Heaven will certainly have perfect environmental conditions, but striving for that in the present is much less important for the fundamentalist than direct proclamation of the gospel. At the same time, there is sufficient basis for a positive environmental ethics within the fundamentalist theological tradition, which can be seen in the proposed theological perspective for environmental ethics.

The evangelical perspective sees human concern for the environment as significant but not essential to Christianity. Caring for creation is important and is an aspect of good stewardship over the created order, but it should be kept in balance with other means of living out the gospel so that it does not obscure the gospel. Within the evangelical category there are a variety of positions on issues such as global warming, industrialization, human rights, and more. There are basic theological presuppositions within this category that tie it together more closely than do particular interpretations of scientific data and the understanding of the role of the local congregation in environmental activism. Although this general approval is shared by many conservative Catholic thinkers, this book focuses on evangelical Protestants because the research has to end somewhere.

There are a million variations between the poles of ecotheology and fundamentalism. Even among people that would describe themselves with these labels, there are going to be debates about doctrines. My point is less to argue that, for example, evangelicals believe certain things, but rather to show that the things people believe about four particular doctrinal questions give a pretty good idea of where their environmental ethics will land. The four types presented in this project are heuristic tools—case studies—used to test the thesis that such a thing as a theological perspective for environmental ethics exists and that the four doctrinal questions are the central points of inquiry for it.

Defense of Typology

Given the dangers of theological labels, it is worth considering why that sort of typology is helpful. One simple answer is that, while imperfect, typology is very helpful in understanding the ongoing debate at a broader level. While Francis Schaeffer's environmentalism may be of particular interest to me, understanding his perspective is only really helpful if it is situated within and, at least to some degree, representative of a particular approach to the topic. Typology can certainly lead to an unhealthy tribalism ("She's just a liberal [fundamentalist].") used to dismiss people without consideration. However, the sneaky defense that some people use by retreating behind the particular ("My view is really closer to Jones, who holds a variant on the environmentalism of Smith, but you've probably never read that.") makes great sense in the conference Q & A (especially from the "more a comment than a question" guy), but it really doesn't advance understanding more broadly or help non-specialist audiences.[19] I'm using a four-fold typology based on broad theological streams in this book to provide a useful set of categories to start with in a field of study with infinite possible variations. As C. S. Lewis once wrote, "Generalities are the lenses with which our intellects have to manage."[20] So, I will risk the accusation of unhelpful reductionism and proceed in hopes of giving someone a foothold to start from.

Categorization is always somewhat reductionistic, but it is also part of the way humans learn. While philosophers and theologians rightly wrestle with whether we have the right categories, cognitive scientists recognize the necessity of categorization for understanding. According to Stevan Harnad, "Categorization is any systematic differential interaction between an autonomous, adaptive sensorimotor system and its world."[21] This is not to say that such categorization is necessarily static. Harnad argues that categories may shift over time due to changes in the cognitive processes of the person receiving the data. We get smarter over time. However, a key recognition is that "categorization is intimately tied

---

19. Armin Geertz wrestles with the use and construction of typologies for the study of the relationship between science and religion. While acknowledging its limitations, he also explains the importance of using types to enable more holistic analysis. Geertz, "Theory, Definition, and Typology," 29–38.

20. Lewis, *Letters to Malcolm*, 74.

21. Harnad, "To Cognize Is to Categorize," 21. Emphasis removed.

to learning."[22] While the topic of categorization in cognitive science is enthralling, a deeper exploration into the topic exceeds the bounds of this project. However, a few observations from the discipline are worthy of note: (1) The ability to understand and process new information is dependent upon the use of categories; (2) Those categories may change over time; (3) Categories are not necessarily absolutely connected to the things they describe, but merely useful for understanding as a heuristic; (4) The lack of static categories does not invalidate the process of categorization.[23]

Heuristic categorization, such as the taxonomic typological system applied in this project, is simply a fact of human understanding and can be made useful by clear definition and explanation. Typology can be useful in simplifying the complex, but, as H. Richard Niebuhr notes, it also has several limitations.[24] First, types are ideal constructs that do not perfectly describe any thinker. Whenever possible, this project explores types in detail using a particular figure to provide concreteness to the abstract. The typological approach of this project is not an assertion that the examples provided are perfect for the type in all respects or that all thinkers who might be associated with a particular type are necessarily "centered" in the type.[25] Second, to maintain a level of cogency, only certain attributes of figures within the types will be considered. It may be true that all adherents of a particular type are relatively wealthy majority thinkers in "first-world" settings, but to combine such observations with the theological elements would stretch the usage of a type.[26] This is not to say that these complicating factors are unimportant, merely that setting them aside may be accepted as a worthwhile exchange in this analysis for the sake of clarity. We still teach Newtonian physics to our children because it is close enough for a beginning understanding.

Furthermore, a typology used to simplify and name for the purpose of examination can provide a structure of categories that cluster similar objects to allow for clearer comparison between writers separated by

---

22. Harnad, "To Cognize Is to Categorize," 22. Emphasis removed.

23. Harnad, "To Cognize Is to Categorize," 19–43.

24. Critics like Don Carson have criticized Niebuhr's typology and called for abandoning it. However, despite these criticisms, the typology retains value and remains useful. Carson, *Christ and Culture Revisited*, 31–43.

25. Hiebert, "Critical Contextualization," 104–12.

26. Niebuhr, *Christ and Culture*, xxxvii–xxxix.

time.²⁷ I use four types as a naming structure simply to test the usefulness of the four key doctrines in a theological perspective for environmental ethics. This approach allows me to engage in critical interaction that rises above the level of mere commentary. In literary criticism, the failure to use taxonomic typology results in what Northrop Frye describes as "quaking bogs of generalities, judicious pronouncements of value, reflective comments, perorations to works of research, and other consequences of taking the large view."²⁸ In lieu of such unhelpful commentary, Frye proposes a systematic, critical approach, a version of which I have attempted to apply here.

The process of categorization in this book has two main axes. The first axis is the structural analysis of the thing itself: in this case the doctrinal questions of a theological perspective. The focus in this axis is the content of a group or particular author's writing on environmentalism. My purpose in writing the book is to argue that focusing on certain significant doctrines (and setting aside others) that are at the center of the issue in question can facilitate meaningful dialogue between theological streams.

The second axis is the application of an external, critical structure: in this project, the categorization of theological streams.²⁹ Among many possibilities, I focus on four different streams of theology: ecotheological, liberal, fundamentalist, and evangelical. Each of these four categories are centered sets with fuzzy boundaries, but even these fuzzy boundaries are necessary to bring coherence to this theological analysis. Sociologists Bernard Zaleha and Andrew Szasz note, "It is helpful to think of Christianity not as one undifferentiated or uniform mass of believers but as an internally divided community."³⁰ They argue, "to understand differences within American Christianity one need[s] to look beyond denominational labels and to think, instead, of three different categories of Christians: One theologically conservative, a second quite liberal, and between them a broad swath of amorphous and ill-defined moderates."³¹ Denomi-

---

27. The importance of an appropriate taxonomic or typological approach is highlighted in Juschka, "Cane Toads, Taxonomies, Boundaries, and the Comparative Study of Religion," 12–23.

28. Frye, *Fables of Identity*, 8.

29. Frye, *Fables of Identity*, 9.

30. Zaleha and Szasz, "Why Conservative Christians Don't Believe in Climate Change," 21.

31. Zaleha and Szasz, "Why Conservative Christians Don't Believe in Climate

national categories may have been cleaner from a labeling perspective, but would have changed the nature of the project to more of a historical analysis, which has its own wrinkles of difficulty. In the end, whether readers agree with my typology of theological streams is not particularly significant. I borrowed three of them from three self-identified ecotheologians and added an evangelical category.[32] The explanatory benefit in this book is in the idea of theological perspectives, not in the accuracy of theological categories.

PROJECT OVERVIEW

At the end of the day, the purpose of this book is to find a way to communicate about environmental ethics between theological tribes. The topic is too important to our day to give up on meaningful communication, because that will just result in the continued escalation of the conflict. There is certainly a lot to disagree about. There is little chance in this life that there will be broad agreement on a wide range of issues. But there can be a degree of cooperation even amidst disagreement. One of my main goals in this book is to frame the shared points of contact in environmental ethics, with the hopes that it will increase the ability to talk to people with differing viewpoints. At the end of the day, everyone wants clean water, clear air, and a future for humanity on this earth. However, to have a conversation, there has to be a shared vocabulary.

The book is divided into five parts. Part One consists of this introduction and one other chapter. Chapter Two will explain in more detail the four doctrinal questions that form a theological perspective for environmental ethics. Those are the common points of contact for a discussion of environmental ethics. Part Two includes an overview of ecotheology in Chapter Three and a deeper exploration of Ernst Conradie's theology for the environment in Chapter Four. In Part Three I outline a liberal environmental ethics and then flesh out the specific ideas of Joseph Sittler's theological perspective for environmental ethics. Part Four has but a single chapter on a fundamentalist environmental ethics, which is a composite built from sources that self-identify or are commonly used by fundamentalists. There was no clear example of a fundamentalist environmental ethics to draw on. Part Five outlines an evangelical theological

---

Change," 21.

32. Horrell, Hunt, and Southgate, "Appeals to the Bible," 219–38.

perspective for environment ethics in Chapter Eight and then digs deeper in the theology of Francis Schaeffer in Chapter Nine. This book closes with a brief conclusion and hopeful exhortation to make progress in this important discussion.

CHAPTER TWO

Four Doctrinal Questions

THERE ARE MANY POSSIBLE approaches used to formulate environmental ethics through a theological lens. For example, a 2014 book, *Systematic Theology and Climate Change*, represents an attempt to do environmental ethics through ten doctrinal headings.[1] In contrast, Whitney Bauman attempts to create a religious approach to environmental ethics through the epistemology of queer theory.[2] Other explicitly Christian treatments of environmental ethics use virtue theory as the basic paradigm.[3] Each of these three projects is theological in the broad sense of the word, but each uses a distinct methodological approach to environmental ethics. The variety in methodology tends to make approaching environmental ethics difficult because, on the surface, there seems to be little continuity between discussion partners. Theological perspective provides a means to dialogue across differing methodologies by identifying a common set of questions.

Throughout the growing body of literature on environmental ethics, there appear to be four questions that are central and common between theological discussions. Those questions are: 1) What are the sources of authority for environmental ethics? 2) Why does creation have value? 3) What is the human role in creation? 4) What is the end goal or final state of the created order and how does it come about? Within Christian

---

1. Northcott and Scott, *Systematic Theology and Climate Change*.

2. Bauman, *Religion and Ecology*.

3. Blanchard and O'Brien, *An Introduction to Christian Environmentalism*; Bouma-Prediger, *Earthkeeping and Character*.

formulations of environmental ethics, these questions can be addressed under the heading of four key doctrines: revelation, creation, anthropology, and eschatology. In non-theological terms, these could fall under different headings, such as epistemology, nature, humanity, and thermodynamics. The boundary of application for these questions need not fall at the edge Christian belief, though that will remain the center of this project.

The four questions, which fall under these four doctrinal headings, tend to be the organizing structure for Christian environmental ethics. These questions are not imposed onto the discussion; they appear to rise from the body of literature on environmental ethics. For example, in a 1992 essay, Millard Erickson identifies four common charges against Christianity:

1. The call to have dominion over the earth, in Genesis 1:28, entails treating the earth as being important only to support the good of the human being. This therefore leads to exploitation and rape of the earth.

2. Christianity has condoned modern science and technology's exploitation of the earth.

3. Christianity has promoted a dualism, according to which the natural or physical or the secular is of less value, or even is negative in character, compared with the spiritual or the otherworldly.

4. Belief in the second coming, which will usher in Christ's complete and perfect reign, effectively removes any reason for us to be concerned about ecology.[4]

Erickson draws these accusations from a number of sources. They can be closely related to the four questions that are proposed as key to a discussion of environmental ethics. The first accusation relates to anthropology and the human place in the created order. The second accusation relates to the role of science in determining what ought to be done; ironically, the complaint that Christians allowed anthropocentric science to encourage the exploitation of the earth has resulted from a new biocentric emphasis within science. In other words, the charge is that Christians

---

4. See Erickson's summary of White's complaint in Erickson, "Biblical Theology of Ecology," 36–37 Note that J. Simmons ascribes these critiques to the doctrinal headings, anthropology, ethics, cosmology, and eschatology. Simmons, "Evangelical Environmentalism," 41n1.

allowed *that* science to inform their actions instead of *this* science. This seems to point to a Kuhnian paradigm shift, where the prevailing culture finds its authority in new science with parts of culture (including, according to the accusation, Christianity) lagging in an older paradigm.[5] This second accusation can be discussed under the heading of the doctrine of revelation. The third accusation relates to a theory of value, which is part of the doctrine of creation, because of the belief that Christianity depends on some sort of neo-Platonistic of gnostic dualism, which requires devaluing the natural world.[6] The fourth accusation relates to the doctrine of eschatology quite directly, though the question remains how significantly a biblical eschatology inhibits positive attitudes toward the environment.

It would be impossible to eliminate from consideration all other doctrines besides these four. In fact, as Northcott and company show in *Systematic Theology and Climate Change*, environmental ethics can be looked at through a number of doctrinal headings.[7] However, a careful reading of the essays in *Systematic Theology and Climate Change* reveals that even as authors are discussing doctrines other than those included in a theological perspective for environmental ethics, they are largely seeking to answer the same four doctrinal questions.

At this early point I will forgive readers for suspecting that I have developed a paradigm that I am in love with and am making all evidence fit that theory, much like the father in *My Big Fat Greek Wedding* who promises to show how any word has its root in the Greek language. He does not bat an eye when a child asks him to explain how the word *kimono* comes from the Greek, but proceeds to pull an etymological connection from the air and expound it with confidence. That is certainly not my intent. Reductionism is not the goal. These four doctrinal questions do not explain everything about environmental ethics, but they do reappear with a regularity that must be more than accidental, which has led me to believe they form the core of the conversation about environmentalism both inside and outside of Christian circles.

---

5. See the helpful summary of Kuhn's theory in Allen and Springsted, *Philosophy for Understanding Theology*, 214–18.

6. E.g., Simmons, "Evangelical Environmentalism," 44–46.

7. Northcott and Scott, eds., *Systematic Theology and Climate Change*. Northcott et al. use ten doctrinal headings, but they all seem to point back to, or at least rely heavily on, revelation, creation, anthropology, and eschatology.

AN EXAMPLE OF THE FOUR DOCTRINAL QUESTIONS

In 2006, E. O. Wilson, Pulitzer prize winner and Harvard biologist, published a volume addressing a hypothetical Southern Baptist pastor. Wilson, who was once a Christian and now professes to be a secular humanist, outlines his basic plea for a Southern Baptist and secular humanist to join forces to save the planet. He argues there are points of similarity between his atheism and Christianity that are sufficient to make common cause for the good of the environment, despite different worldviews. To be clear, while Wilson mentions salvation through Christ, this is only a tangential reference; Wilson is concerned with the environment alone and not with the eternal state of human souls.[8] Wilson does not enumerate a list, but his introduction shows there are four main points of contact—four key doctrines—between his worldview and the Southern Baptist's that must be considered for communication to take place.

Wilson begins by describing the pastor as "a literal interpreter of Christian Holy Scripture."[9] For Wilson there is an obvious tension in sources of authority between science and Scripture. He notes the pastor seeks special revelation while he, the biologist, is satisfied with "the glory of the universe revealed at last."[10] Wilson argues that both he and the Christian can find universal value in nature. For Wilson, this is because "it serves without discrimination the interests of all humanity."[11] This is the doctrine of creation for Wilson. He argues the pastor should desire to save the earth because "each species . . . is a masterpiece of biology, and well worth saving."[12] There is, then, a degree of value for nature in the mind of Wilson on both instrumental and aesthetic grounds.

In this brief introduction, Wilson also argues for an evolutionary view of humanity.[13] This is to be expected, but it has implications by placing humans within the web of the ecosystem. There is, then, for Wilson no distinction between humanity and the rest of creation. For Wilson this is a point that reflects both the authority of science over the discussion and the role of humans within nature. The fourth point that Wilson highlights significantly in this supposed letter is that of eschatology, though

8. Wilson, *The Creation*, 3.
9. Wilson, *The Creation*, 3.
10. Wilson, *The Creation*, 4.
11. Wilson, *The Creation*, 4.
12. Wilson, *The Creation*, 5.
13. Wilson, *The Creation*, 4.

he does not use the term. According to Wilson, the vision of a destruction of the created order some Christians hold results in apathy toward the creation, which he is trying to overcome.[14]

Having pleaded for common ground, Wilson then goes on to abandon his common ground approach as he provides a scientific appeal for saving the planet. After recognizing the major points of contention between his view and his purported audience, he proceeds to shift away from the discussion of the differences and make his case on his terms.[15] In his recommended solutions for developing children that care about preserving nature (or creation), he instructs his Southern Baptist friend to teach children "that all biological processes, and all the differences that distinguish species, have evolved by natural selection."[16]

Wilson recognizes there are differences in understanding, but then he asks his audience to deny their core beliefs and accept his. There can be common ground as long as everyone agrees with Wilson. He is unlikely to appeal to his audience because he fails to show how individuals with different perspectives on the four key doctrines related to environmental ethics can work together. He also fails to argue for a change in positions. He simply assumes the reader will adopt his assertions. *Creation* seems more designed to appeal to Wilson's base than as a genuine outreach to Southern Baptists.

Whatever disagreements remain, as a secular humanist, Wilson demonstrates that there are discernable touch points of concern for environmental ethics within Christianity. His plea is evidence of recurring patterns of doctrines that appear to influence environmental ethics more significantly than others. In fact, as can be seen in Wilson's introduction, there are four core key doctrines that form a theological perspective that shapes environmental ethics. Unfortunately, to date there has been little to no work done on methodology within environmental ethics. This is not altogether surprising since the discipline is only a few decades old. However, this lacuna makes the beginning process of evaluating

---

14. Wilson, *The Creation*, 6.

15. For example, in his later discussion of human nature and the desire to conquer the environment, Wilson describes humans as little more than biological machines, dependent on chemical impulses that were wired through heredity. While this is consistent with his view, it does not seem calculated to convince a Southern Baptist pastor. Wilson, *The Creation*, 64.

16. Wilson, *The Creation*, 113.

environmental ethics exceedingly difficult. What is being offered in this book is a method to help make sense of a very complex field of study.

EXPLAINING THE FOUR DOCTRINAL QUESTIONS

If readers accept the argument that the four doctrinal questions are central, if not exclusive, to debates over environmental ethics, then it still remains for the questions to be more precisely shaped. Each of the four major doctrines has behind it a vast volume of scholarship. Much of that scholarship uses different terminology. This section provides a narrowed definition of each of the four doctrinal questions, categorized under a doctrinal heading, that will undergird the arguments in each of the chapters of this volume.

Revelation

The first task in any theological debate should be to establish the nature of various sources of authority. Every Christian environmental ethicist must determine what forms of revelation are authoritative and to what degree they inform theological method. The so-called Wesleyan Quadrilateral provides a useful model for considering sources of authority. Thus, this project evaluates the balance between reason, Scripture, experience, and tradition as part of the characterization of each of the theological types.[17] The discussion of authority among Christian environmental ethicists typically focuses on the value of Scripture and the normative force of science, science being largely a conflation of reason and experience. Apart from its contribution to observational science, experience in conservative theological streams can be a motivator for environmental action, rather than a source of authority. For liberal environmentalists and ecotheologians, experience tends to be a significant source of authority for directing environmental ethics. Tradition plays a small role in most theological perspectives for environmental ethics, largely because the environmental discussion is very recent in Christian history.

---

17. John Wesley never used the term "quadrilateral" in his own writing. The term was coined by Albert Outler and included in the 1972 UMC *Book of Discipline*: Thompson, "Outler's Quadrilateral, Moral Psychology, and Theological Reflection in the Wesleyan Tradition," 53n15. Outler himself outlines the so-called Wesleyan Quadrilateral in an article in which he admits to regretting coining the phrase.

Understanding the authority of Scripture within a theological perspective for environmental ethics is of first importance for Christians. The degree of authority granted to Scripture, the reliability of Scripture, and the function of Scripture in theological method are all formative for an environmental ethics. Robert Fowler argues, "The best guide to the contours of the division [between Protestants over the environment] appears to be knowing how much a given Protestant believes the Bible is absolutely true. The more 'fundamentalist' Protestants are in this sense, independent of other variables, the lower they will be, relatively, on assorted indices of environmentalism."[18] He also notes that for many Protestant environmentalists, "Whether this fact is acknowledged or not, science often appears to be the ultimate authority."[19]

The reason and experience legs of the Wesleyan quadrilateral can be largely discussed in terms of science for this project. Scientific reasoning as a source of moral authority is closely tied to the doctrine of general revelation. If general revelation is considered by an ethicist to be a viable source for ethical reasoning, that will enable a different theological perspective for environmental ethics than one that is more suspicious of human reasoning and experience. So then, the debate over authority for environmental ethics often comes down to a balance between the authority of science and religion.

For most ethicists, there seems to be no hard division between religion and science. Indeed, for most Christians, all truth is God's truth. Therefore, many Christians resonate with the arguments, similar to the one of Abraham Kuyper, that science should be a primary discipline of Christian interest because "it brings to light the hidden glory of God; it gives you joy in the act of digging up the gold that lies hidden in creation; and it grants you the honor of raising the level and well-being of human life."[20] The question is not, therefore, whether there is truth to be found in scientific research, but how fully the data discovered can be trusted for moral decision making.

A complete treatment of the debate over the relationship between science and religion is beyond the parameters of this project. It is important, however, to note the significance of the tension between the

---

18. Fowler, *The Greening of Protestant Thought*, 25.

19. Fowler, *The Greening of Protestant Thought*, 5.

20. Kuyper, *Scholarship*, 9. Note that the word translated "science" here is broader than the contemporary word allows and includes a world-interested scholarship. Kuyper, *Scholarship*, 4n5.

two because it is a constant subject of dialogue within contemporary theology. Roger Olson argues, "Much modern theology is dominated by overt or covert attempts to respond to the scientific revolution."[21] Olson spends more than seven hundred pages demonstrating how science has shaped religion in the modern era. The influence of science impacts environmental ethics, as well, especially when the subject is approached as moral theology. According to Christopher Clausen, the position of science as authority is a vital component of any ethics, in large part because much of the scientific community is significantly biased toward political liberalism and against theistic belief. Clausen argues that even supposedly objective science begins to develop moral force in the absence of a theistic framework.[22] The presuppositions of supposedly value-neutral science may bias the interpretation of general revelation as it is applied to environmental ethics.

There are four prevailingly common models of the relationship between religion and science. The independence model sees the two disciplines as entirely separate in their objects, aims, and methods. This is a common perspective among the Neo-Orthodox, existential philosophers, positivists, and those influenced by Wittgenstein.[23] A second model, the conflict model, asserts that religion and science overlap in their object, aim, and method. Accordingly, "religious fundamentalism seeking to refute scientific claims on theological grounds falls into this category."[24] On the other hand, scientist Freeman Dyson argues, "The common element [in a scientific vision of the world] is rebellion against the restrictions imposed by the locally prevailing culture."[25] His comments demonstrate that the perceived conflict arises from both directions. A third model is that there is a dialogue between science and religion; there are differences in method and scope, but the two overlap and are mutually influential. The fourth model of relationship between science and religion is that of integration.[26] In this vision of the relationship, "Proponents of a 'theology of nature' embrace current science to reformulate and refine theological doctrines (e.g., creation, providence, and human freedom)—so that

---

21. Olson, *The Journey of Modern Theology*, 44.
22. Clausen, "Left, Right, and Science," 16.
23. Peterson et al., *Philosophy of Religion*, 509–10.
24. Peterson et al., *Philosophy of Religion*, 510.
25. Dyson, "The Scientist as Rebel," 1.
26. Peterson et al., *Philosophy of Religion*, 510.

science inevitably affects our understanding of God-world interaction."[27] These four models of engagement between science and religion do not map directly onto the four theological types being used in this project. However, some models are more prevalent in certain theological types. For example, the conflict model is much more likely among theological fundamentalists, while the integration model is much more likely among ecotheologians. As such, the models will be used descriptively in discussions, and this explanation of options is provided to inform those descriptions.

Particularly in a Protestant Christian context, the debate between science and religion becomes more directly focused on the tension between Scripture and science, since the principle *sola scriptura* is one of the chief battle cries of the Reformation. In its caricatured form, *sola scriptura* becomes the sort of biblicism that rejects all ideas that appear to disagree with the content of Scripture. From this position, argues Olson, "to deny that God created the entire universe in 4004 BC [is] tantamount to denying the authority of the Bible."[28] However, historical evidence indicates this description of Christian attitudes toward science is less than accurate outside of fundamentalist circles.[29]

There is a constant dialogue in every theological method about the usefulness of scientific data for making ethical choices.[30] This is unavoidable in any discussion. Sometimes bad science results in bad ethics. However, in the end, a sound theological method allows for observations based on bad science to be corrected or reconsidered when a better scientific understanding is achieved.[31]

---

27. Peterson et al., *Philosophy of Religion*, 511.

28. Olson, *The Journey of Modern Theology*, 41.

29. Stiling, "Natural Philosophy and Biblical Authority in the Seventeenth Century," 115–36.

30. The debate over the comparative authority of science and Scripture for ethics is just a part of a larger discussion on the nature of revelation and the authority of Scripture. For example, Roman Catholic scholar Matthew Levering has published an excellent volume dealing with the various understandings of the role of the church (both in the sense of tradition and present community) as revelation. In Levering's model, Scripture and the church are intermixed as one (largely) coherent stream of revelation. Levering, *Engaging the Doctrine of Revelation*, 15–16.

31. For example, Augustine's references to observations from his day, which were "scientific" in a sense, are humorous, but the errors do not diminish the overall value of his theology for having been wrong. Augustine, *The City of God*, XVII.7–9.

The fourth leg of the Wesleyan Quadrilateral is tradition. In environmental ethics, the role of tradition tends to be in deliberating the authority of Scripture for moral theology. Environmental ethics is a new topic for concern, at least to the degree that it presently is discussed. As a source of authority within Christian ethics, tradition is limited. Conservatives appeal to tradition for evidence of the centrality of Scripture in all of Christian theology. Revisionists of various degrees tend to reject Christian tradition as anthropocentric and bad for the environment, except where elements can be retrieved and repurposed to support the environmental agenda.[32]

Creation

In discussing the doctrine of creation for environmental ethics, the value of the created order is much more significant than apologetic arguments about the relative timing of the creation of earth. Different interpretations of the cosmology of Genesis do have significance in the formulation of a theological perspective.[33] However, for environmental ethics, the doctrine of creation is important mainly because it defines a theologian's view of the goodness of creation and the relationship between the current state of the created order and an Edenic or eschatological state.[34]

Some theologians argue that an evolutionary understanding is necessary for a proper environmental ethics and that an affirmation of any form of theistic creation undermines environmentalism. That conclusion is not warranted.[35] Instead, perspectives on the age of the earth tend to reflect a particular relationship between the disciplines of religion and science, which is folded into the discussion of revelation.[36] Arguments

32. E.g., Conradie, "Revisiting the Reception of Kuyper in South Africa," 15–54.

33. Although some argue a young-earth creationist position is required to claim a belief in inerrancy of Scripture, it is doubtful that this is, in fact, necessary. Snoeberger, "Why a Commitment to Inerrancy Does Not Demand a Strictly 6000-Year-Old Earth," 3–17. For a philosophical discussion of this debate, see Smith, "Secularity and Biblical Literalism," 205–19.

34. Wirzba, *From Nature to Creation*, 20–24.

35. Eaton, "The Revolution of Evolution," 6–31. Woodrum and Hoban, "Theology and Religiosity Effects on Environmentalism," 193–206.

36. The science vs. religion debate is one that seems to pit political conservatives against political liberals. This is often framed as a debate over creation vs. evolution. However, Christopher Clausen argues that the main historic figure associated with creationism, William Jennings Bryant, was actually "the farthest-left presidential

about the timing and method of the earth's origin tend to distract from the more significant discussion of the value of the created order.

In a 2013 article, R. J. Berry argues that the creationists' rejection of evolutionary theories results in dualism and a destructive attitude toward the environment.[37] According to Berry, the creation-evolution debate is a stronger cause for the failure to act on environmental data than even the relative priority of evangelism over good works in the world. His conclusion appears to exceed the warrant of evidence; however, his argument is not without merit, since fundamentalist theologians tend to treat the doctrine of creation mainly under the heading of anthropology and focus on a defense of creationism instead of explaining the value of the created order.

Notably, Berry calls for the acceptance of evolutionary theory as unquestionably factual. He writes, "Religious debates over evolution are significant for students or seekers (as Augustine realised), but they are much more important in distracting from the general need to understand God's relation to his creation. Christians will continue to ignore the need and responsibility for creation care if they continue to argue about evolution."[38] Although Berry is most scornful of those who adhere to a young-earth creationism, mainly fundamentalist Christians, he is similarly dismissive of Christians that hold to other theories of origin, such as intelligent design (I. D.).[39] Berry makes the ramifications of his view plain, stating, "Natural theology in the Thomist sense as proof of God's existence is dead."[40] Concern over seeing God's handiwork in creation distracts from the work of environmentalism. In fact, in a footnote, Berry states that interviewing advocates of I. D. leaves Lee Strobel's book, *The Case for a Creator*, "fatally marred."[41]

---

nominee in US history . . . If the South had not been simultaneously more religious and more conservative (for unrelated reasons) than the rest of the country when these controversies came to a head, Christian belief might easily have been more often identified with liberal politics and evolution with the right." Clausen, "Left, Right, and Science," 16. Whether Clausen's assertions are correct is debatable, but it does point to the difficulty of blending religious and political positions or conflating particular understandings of religion with particular understandings of science.

37. Berry, "Disputing Evolution Encourages Environmental Neglect," 128–29.
38. Berry, "Disputing Evolution Encourages Environmental Neglect," 120.
39. Berry, "Disputing Evolution Encourages Environmental Neglect," 119.
40. Berry, "Disputing Evolution Encourages Environmental Neglect," 120.
41. Berry, "Disputing Evolution Encourages Environmental Neglect," 119n24.

Berry argues that the understanding that God was intimately active in the creative process results in dualism. According to Berry, the effect of the efforts of creationists (in which group he includes those who support I. D.) "is that creation is devalued from being God's world as described in the Bible; it is reduced instead to a 'thing,' one brought about by divine fiat but where the sense of the intimate involvement of God in both the past and present outcomes of the world so vividly described by evolutionary biologists is completely lost."[42] The apparent reason given for the rejection of God's active work in continued creative activity is rooted in the sharp division Berry makes between God's work as creator and as sustainer.[43]

In the end, Berry's more general point seems valid. Many Christians fail to move past debates about cosmology, which contributes to a weak understanding of the value of creation. However, his argument that it is necessary to accept the evolutionary theory as many scientists commonly explain it does not follow. In fact, Berry shows that a positive environmental ethics can be developed by reading Scripture with a high view. Such a constructive theological enterprise does need to take place among fundamentalists, but it does not appear to require a rejection of traditional understandings of God's work in creation.

Value of Creation[44]

In environmental texts, value is generally discussed as being either *intrinsic* or *instrumental*. Intrinsic goodness entails the goodness being native to an object itself for its own sake.[45] In contrast, instrumental value is the utility of an object to a subject.[46] For environmentalists, viewing nature as having intrinsic value is considered good, while ascribing instrumental value to the created order is generally considered bad.[47] Neither of these

---

42. Berry, "Disputing Evolution Encourages Environmental Neglect," 128.

43. Berry, "Disputing Evolution Encourages Environmental Neglect," 119.

44. Portions of the discussion on the value of creation are based on the research of a previously published article: Spencer, "The Inherent Value of the Created Order," 1–17. My thanks go to Robert George, editor of *Theoecology Journal*, for his kind permission to re-use portions of that article here.

45. Clarence Lewis defines intrinsic value as "that which is good in itself or good for its own sake." Lewis, *An Analysis of Knowledge and Valuation*, 382.

46. Lewis, *An Analysis of Knowledge and Valuation*, 391.

47. E.g., Callicott, *In Defense of the Land Ethic*, 130–31.

categories of value, however, seem sufficient for describing the value of the created order that is due to a relationship to an external being, but not dependent on the utility of the created order to that external being.[48] For this project, a third category of value, inherent value, is being proposed.[49] Inherent value describes the goodness of the created order based on the proper orientation to that which imparts value to it.[50] In this case, given the Christian presuppositions of this project, God is the being who established the inherent value of the created order.[51]

Toward a More Nuanced Theory of Value

The topic of value theory is complex. Even between systems of value that seem to agree in principle, there are differences in vocabulary that can confuse the discussion. It is possible to understand value through a biblical worldview and describe the schema of value in significantly different ways. Finding a key to the important ethical category of value can be difficult for a Christian, particularly since many philosophical systems that discuss value seem to disallow the existence of God. The three-part value theory proposed in this section is necessary for clarity in this project, however unlikely it is that the framework is adopted universally.

Augustine provides an early Christian foundation for value theory. In *The Nature of the Good*, Augustine argues God is good in a unique

---

48. J. A. Simmons notes discussion around the use of "intrinsic" as the category of valuation of the created order for evangelicals in particular from David Wood, an apologist and philosopher. Simmons acknowledges the need to qualify the notion of intrinsic value, but decides to use the descriptor anyway. Simmons, "Evangelical Environmentalism: Oxymoron or Opportunity?," 46–47

49. Robert Nelson notes the insufficiency of the term "intrinsic" as it is commonly used: "Something cannot have a 'value'—for a value cannot exist in a vacuum—unless there is a valuer, human or otherwise." Nelson, "Calvinism without God," 259. He goes on to define a sort of value which he calls intrinsic, but it is not self-contained; it is inherent as the category has been used in this project.

50. According to Lewis, inherent value is a subset of extrinsic values. Lewis defines inherent values as "those values which are resident in objects in such wise that they are realizable in experience through presentation of the object itself to which they are attributed." Lewis, *An Analysis of Knowledge and Valuation*, 391.

51. In an essay on literary criticism, Leland Ryken notes, "The Christian doctrine of creation carries with it a respect for the integrity of created things. Things have inherent value because they are part of the created order. A tree has value as a tree because that is what God created it to be. Its value depends on its fulfilling its created purpose." Ryken, "Formalist and Archetypal Criticism," 16.

way. Since God is the creator, he is distinct from the creation.[52] Thus his goodness is a higher goodness than that possessed by created things. Creation is good, but it has a goodness derived from its relationship to the Creator.[53] The degree of derived goodness of an object is determined by its fulfillment of the purpose for which God designed it.[54] This is inherent value.

Augustine's basic value structure resonates with a commonly accepted division in value theory between intrinsic and extrinsic goods.[55] There is an immutable, final value that Augustine attributes to God alone, which is intrinsic.[56] Extrinsic goodness is non-final value ascribed to an object based on its relationship to a set of qualities.[57] Thus, extrinsic value refers to goodness that is due to attributes that come from outside the object itself. Outside of Christendom, Harvard scholar C. I. Lewis, a non-theistic pragmatist philosopher, uses a category for intrinsic value, but argues, "All value in objective existents is extrinsic."[58] Lewis sees two categories but argues intrinsic value does not exist in reality. Such an assertion makes sense for a non-theist, because an object with independent value would seem to have the attribute of aseity, which Christians ascribe to God alone. Thus, Lewis's understanding of intrinsic and extrinsic value roughly matches an Augustinian system and can be applied in a Christian context.

Limiting the categories of value to intrinsic and extrinsic is insufficient because it conflates differing types of extrinsic value, which can hamper later ethical discussions.[59] At least two types of extrinsic good-

52. Augustine's Creator-creature distinction has led to accusations of dualism from some: Pagels, *Adam, Eve, and the Serpent*, 99; Gunton, *The Triune Creator*, 79. However, Bradley Green has since artfully corrected Gunton by arguing that Augustine is a hierarchical dualist, thus affirming the Creator-creature distinction without devaluing the created order. Green, *Colin Gunton and the Failure of Augustine*, 132.

53. Augustine, "The Nature of the Good," 1.

54. Augustine, "The Nature of the Good," 8.

55. Ronnow-Rasmussen, "Intrinsic and Extrinsic Value," 29–30.

56. Augustine, "The Nature of the Good," 1.

57. Ronnow-Rasmussen, "Intrinsic and Extrinsic Value," 32.

58. Lewis, *An Analysis of Knowledge and Valuation*, 432. Lewis's main goal was to combat the influence of skeptical philosophies by arguing for a rational, objective ordering in the world. This is why Lewis is helpful for developing a Christian value system, though he is not himself a believer. Colella, "Human Nature and the Ethics of C. I. Lewis," 302.

59. Ronnow-Rasmussen, "Intrinsic and Extrinsic Value," 29.

ness are indicated. The first type of extrinsic values are inherent values, which are, according to Lewis, "those values which are resident in objects in such wise that they are realizable in experience through presentation to the object itself to which they are attributed."[60] Goodness of an object is thus due to qualities in the object's relationship to some external set of values. A second type of extrinsic values are instrumental values, which are dependent on the utility of an object to a subject.[61] In the instrumental view, goodness of an object is dependent on how well it fulfills a desired end.

There is both distinction and connection between categories of instrumental and inherent value. For example, a painting may be beautiful and representative of artistic excellence but serve no practical purpose. The painting would have inherent value, but little instrumental value.[62] The value of such a painting is primarily derived from its ability to delight the viewer and testify to the excellence of the painter. Its value is relational. In contrast, a dust mop may have mainly instrumental value, inasmuch as it is good for cleaning the floor of the art gallery. However, if it is leaned next to the painting by a careless custodian, there would be no comparison between the beauty of the two objects, and it would give little or no attestation to the artisanship of the manufacturer.[63] This is especially true for objects that have been hastily and inexpensively made with only utility in mind. On the other hand, a staircase in the same art gallery may have significant instrumental and inherent value simultaneously if it is both artistically excellent and useful for transiting between levels. Instrumental and inherent value are largely independent categories from one another.

Augustine's concept of goodness is helpful for discussing intrinsic value and both types of extrinsic value, as well. In his discussion of the use and enjoyment of things in *De Doctrina Christiana*, Augustine differentiates between objects that are to be "used" and those that are to be "enjoyed." Those things that are to be enjoyed have intrinsic value, "For

---

60. Lewis, *An Analysis of Knowledge and Valuation*, 391.

61. Lewis, *An Analysis of Knowledge and Valuation*, 391.

62. It might be argued that even the painting would have some utility if it were used to momentarily feed a fire or stop the draft around a door. This, however, is not the primary purpose of the painting and would reflect little inherent value.

63. One might argue that the dust mop might properly attest to the quality of a brand and thus have some inherent value. The limited relation to its maker reflects a minimal inherent value.

to enjoy a thing is to rest with satisfaction in it for its own sake."[64] This underived value is attributed to God alone.[65] Created objects, which have inherent value inasmuch as they are rightly ordered to God's design, are only to be used. As Oliver O'Donovan writes, "Augustine's position . . . is that one 'uses' an object that is in itself 'for use' (*utendum*) and enjoys an object which is in itself 'for enjoyment' (*fruendum*). To break with the objective order in which this distinction is rooted is vicious and perverse."[66] The Creator is to be enjoyed and the creation is to be used. To do otherwise is abuse, which is sin.

In contemporary discussions, the term "use" generally has a negative connotation, reflecting only instrumental value. This is particularly troublesome when the use of humans is considered. However, as O'Donovan explains, "Our 'use' of other people must be understood to promote our welfare as well as theirs."[67] Thus, using an object, human or otherwise, in Augustine's terms indicates that it must be appropriately valued in light of its relationship to God, and the use of it should enhance that relationship rather than detract from it. To use an object is to recognize and honor its inherent value.

Inherent value belongs to objects in proportion to the degree to which they are properly related to God. There is still goodness in things with inherent value, but it is not the highest good, and it is derived from the ordering of the nature of the object to God. This reflects the distinction between Creator and creature. These lesser goods should be used (in the Augustinian sense) and not enjoyed.

Objects with inherent value may be used to glorify God, who is the only object with intrinsic value, and the only object that should be enjoyed. As Augustine writes about creation in *Confessions*, "Your works praise you that we may love you, and we love you that your works may praise you."[68] God's works have inherent value and may also have instrumental value. In both cases they may be used or abused according to Augustine's definitions.[69] However, an object may have instrumental value

---

64. Augustine, *On Christian Teaching* 4.4.
65. Augustine, *On Christian Teaching* 5.5.
66. O'Donovan, "Usus and Fruitio in Augustine, *De Doctrina Christiana I*," 368.
67. O'Donovan, "Usus and Fruitio in Augustine, *De Doctrina Christiana I*," 371.
68. Augustine, *The Confessions* 13.33.
69. Indeed, in his *Responses to Miscellaneous Questions*, Augustine writes, "Everything that has been created, then, has been created for the use of human beings, because reason uses with judgment everything that has been given to human beings."

despite being deficient (though not entirely lacking) in inherent value. Augustine supports this when he shows how God used Satan to prove Job's righteousness.[70] Thus Satan's disordered being, with its diminished goodness, has instrumental value to God greater than his inherent value. There is a difference between these two lesser types of value. Likewise, an object improperly used is being abused, as the staircase from the example above would have been abused if it was employed for committing murder by pushing someone down it.[71] It may have instrumental value for that improper use, which would be divorced from the inherent value connected to its ordered usefulness. Its instrumental and inherent value is misaligned in an instance when the staircase is misused. Goodness is inherent to the created order because it was created by the highest good and its inherent value is proportional to its ordering to God, but the goodness of the created order is also instrumental, as it may be used or abused.

Value Theory for Environmental Ethics

This three-category system of values has support from non-Christian philosophical streams, relates to an Augustinian understanding of ordered goods, and seems to have benefits for developing a balanced view of the goodness of the created order. The three categories—intrinsic, inherent, and instrumental—are not universally accepted within environmental ethics. However, they do provide clarity of vocabulary for a cross-categorical discussion of the value of creation. Using the three-part value structure outlined above is helpful for environmental ethics, because there is a great deal of confusion among environmental ethicists about the true nature of value.

Ecotheologians tend to ascribe intrinsic value to nature. Such an attitude lends itself toward pantheism or panentheism, which is a common theme among radical environmentalists. While the intent of ascribing intrinsic value to nature appears to be rejection of the mere instrumental valuation of the created order, valuing creation intrinsically renders environmental decision-making impossible.[72] In colloquial terms, if everything is important, then nothing is important.

Augustine, *Miscellany of Eighty-Three Questions*, 30.
    70. Augustine, "The Nature of the Good," 32.
    71. Augustine, *On Christian Teaching* 4.4.
    72. Sarkar, *Biodiversity and Environmental Philosophy*, 57.

Liberal theologians also have a tendency toward pantheism or panentheism.[73] Thus, when theologians in the liberal theological stream describe creation as intrinsically valuable, there is a strong possibility of a certain amount of deification of the created order. At the same time, other liberal ethicists, like James Gustafson, use the term intrinsic and define it in the way inherent value is described above.[74]

Evangelicals tend to struggle with describing the value of creation. Francis Schaeffer, the first major evangelical thinker to write on environmental ethics, uses the word "intrinsic," but clearly utilizes it to mean "inherent" as the term is defined in this project.[75] Still other evangelicals, like Liederbach and Bible, use the terms interchangeably but mean "inherent," as defined here.[76] Meanwhile, fundamentalists tend to discuss only the instrumental value of creation, though they rarely deny other forms of value. Such definitional confusion illustrates the need for the framework provided in this section for clarity in comparing different approaches to environmental ethics.

Anthropology

The inclusion of anthropology as one of the four key doctrines in a theological perspective for environmental ethics seems intuitive to me, since the contemporary ecological debate largely hinges on the human-environment interaction; this interaction is inherently anthropological. However, it is one of the places that I have gotten the most pushback in discussions on the framework. The objection is that by addressing anthropology directly, I risk making environmental ethics anthropocentric. I cannot deny that the risk is there, but there are two major reasons why it is vital to address anthropology when writing about environmental ethics. First, because there is a strong anti-human element in contemporary environmental discussions. A more extreme example comes from ahumanist philosopher Patricia MacCormack, who states, "Human extinction can be understood as a good idea for ecosophical ethics and need not be considered 'unthinkable' but can be welcomed as affirmative of

---

73. Olson, *The Story of Christian Theology*, 550.
74. Gustafson, *A Sense of the Divine*, 55–58.
75. Schaeffer, *Complete Works*, "Pollution and the Death of Man," 5:34.
76. Liederbach and Bible, *True North*, 35–50.

earth life."[77] Clearly the goodness of humanity needs some defense when discussing the environment. Second, the doctrine of anthropology must be discussed in environmental ethics because the limits of humanity's authority over creation must be understood. There are too many voices on the extremes in their understanding of the role of humanity within the created order to ignore the question entirely.

Under the heading of anthropology, we will consider the nature and impact of the fall on humanity and the relationship between human and non-human creation. The question of the proper relationship between humans and the created order is how even naturalistic theologies move from "is" to "ought"; even those that recognize all revelation as coming through nature will find some duty for humans to interact differently (or less) with the world around them than is customary in the developed world.

Often anthropological views are characterized by different metaphors for environmental ethics.[78] Fundamentalists are sometimes described as having a domination view, which argues for the superiority of human concerns to environmental concerns by such a degree of magnitude that little consideration need be given to a theological approach to environmental ethics. In a more favorable light, however, the fundamentalist understanding of the human role of creation is one of ownership granted by God, with a right to use creation, even to use it up. Conservative evangelicals reflect a dominion ethic (more often labeled "stewardship"), which understands humans to have a role as vice-regents in tending the created order on God's behalf. Liberal theologians tend to see humans as servants of the created order, dependent upon it and dedicated to improving it. Ecotheologians see humans as an alien species from which the created order must be freed by cooperation between humans and a liberating God. Each of these four descriptions reflects a different understanding of the place of humanity within the created order.[79]

The inclusion of this small segment of the doctrine of anthropology—understanding the human place within the created order—should not be a controversial part of the environmental discussion. Proponents from every theological perspective recognize the significance of human-creation interactions for environmental ethics. Additionally, even those

---

77. Patricia MacCormack, *The Ahuman Manifesto*, 144.

78. Donald McDaniel, "Becoming Good Shepherds."

79. The general statements about each of the theological streams will be supported in Chapters Three through Six.

that see a greater right for humans to change the environment for the benefit of humans see that, at times, humans have abused the created order. There are limits to what is good for humans to do. The difference from other streams is that for those Christians within the fundamentalist theological category, spoiling nature tends to reflect bad stewardship and wastefulness rather than a violation of a right of nature to exist. Answering the basic question of the human's proper relationship to non-human creation by no means represents a fully-orbed anthropology, but it gets at the marrow of anthropology's influence on environmental ethics.

Eschatology

The final doctrine of particular concern in a theological perspective for environmental ethics is eschatology. There is much more consensus on the importance of this doctrine for environmental ethics than other doctrines, and it is frequently referenced as central to the neglect of the environment by fundamentalists.[80] Geographer Janel Curry-Roper argues, "I believe that eschatology is the most ecologically decisive component of a theological system. It influences adherents' actions and determines their views of mankind, their bodies, souls, and worldviews."[81] As the nature of creation affects the Christian's view of environmental ethics, so does the ultimate destiny of the created order. A theological perspective for environmental ethics must therefore consider the questions of the nature of the end state and of the means of arriving at the end state of the created order.

According to some environmentalists, the problem with so-called anti-environmental Christians is their eschatological view of the created order, which opponents describe as a vision of creation as a temporary construct that will pass away in a fiery judgment.[82] The view here is total destruction with subsequent entirely new creation, which is typically ascribed to dispensational fundamentalists. As Norman Wirzba demonstrates, however, it is difficult to find evidence of this popularly-held view in academic writing.[83] Some evangelical believers share a destruc-

---

80. For example, Curry-Roper, "Contemporary Christian Eschatologies and Their Relation to Environmental Stewardship," 157–69; Skrimshire, "Eschatology," 162.

81. Curry-Roper, "Contemporary Christian Eschatologies and Their Relation to Environmental Stewardship," 159.

82. Zaleha and Szasz, "Why Conservative Christians Don't Believe in Climate Change," 24–25.

83. Wirzba, *From Nature to Creation*, 21.

tionist view with fundamentalists, but many evangelicals expect a fiery purification of the created order with a subsequent divine renewal of all things in the created order through a special divine work. This latter understanding of eschatology is discussed in the evangelical chapter. The liberal theological perspective tends to present the idea that things will continue roughly as they are, with little emphasis on a divine intervention in the eschaton. The final view—that of ecotheology—is that the eschaton is largely figurative. Eschatology, therefore, provides a metaphor for the need for liberation, and the hope of liberation at the end of days. Although the understandings of eschatology differ wildly, the importance of the doctrine for environmental ethics is apparent in all four theological types.

The significance of eschatology is driven, in part, by the anticipated fate of the created order. However, the impact goes deeper than whether there is an annihilation of the present earth or simply a purification. Eschatology impacts interpretation of history and projection of the future. For some dispensationalists, there is an inevitable negative trajectory for history. Once God's timeline is complete, Christ will come again. In the meanwhile, the task for the Christian is to share the gospel. In at least some versions of this position, there is an idea that when the last predestined human makes a profession of faith, then the rapture will happen and the tribulation will begin. This sometimes results in an emphasis on evangelization at the expense of other good works. Other eschatologies tend to be more hopeful, seeing a future redemption of all things, which means there is some direct benefit in working for the common good in the time before the eschaton.

Each of the four theological streams considers eschatology very important, but there is a much wider variation within and between theological streams on this topic than the other three doctrinal questions, resulting in more overlap between the theological streams. However, in general, distinct understandings of eschatology shape the way ethicists from different theological streams view the created order and the role humans have in tending it. It is generally considered a vital part of the theological system that undergirds all ethics, and it consistently appears as one of the core doctrines for environmental ethics.[84]

---

84. Curry-Roper, "Contemporary Christian Eschatologies and Their Relation to Environmental Stewardship." Though he does not focus on environmental ethics specifically, Braaten argues eschatology is the central doctrine in all ethics. Braaten, *Eschatology and Ethics*, 105–22.

PART TWO

ECOTHEOLOGY

CHAPTER THREE

An Ecotheological Perspective for Environmental Ethics

This chapter examines ecotheology, which is an approach to environmental ethics that emphasizes setting creation free from bondage caused by human sin, through the redefinition of traditional Christian doctrines. On its surface, the term "ecotheology" seems to be merely a handy conflation of the words *ecological* and *theology*. Early on in the history of Christian environmental ethics, the word was used in this fashion, but in most recent Christian scholarship, "ecotheology" has a specialized meaning that implies a specific theological method.

This chapter defines the nature of ecotheology, arguing it is a fourth stream of liberation theology and outlining the origin of and shift in the use and meaning of the term "ecotheology." Then it explains the main theological emphases of ecotheology along the four doctrinal headings of a theological perspective for environmental ethics.

DEFINING ECOTHEOLOGY

In his book *The Journey of Modern Theology*, Roger Olson describes three main streams of liberation theology. Olson argues that liberation theology as a distinct movement has its origin in 1968 when Latin American Catholic hierarchy called for the church to align itself with the poor majority of the world.¹ The original shape of liberation theology can be seen clearly in the work of Gustavo Gutiérrez. His book, *A Theology of Liberation*, published in 1971, remains a trusted introduction to liberation theology.² The rise of Latin American liberation theology was soon melded with a feminist liberation approach, which is well represented by Mary Daly and Rosemary Radford Ruether. At roughly the same time, James Cone developed his Black theology, publishing *A Black Theology of Liberation* in 1970.³ These are the three streams Olson outlines in his account of liberation theology. In *A Survey of Recent Ethics*, published in 1982, Edward Long reveals these three streams of liberation theology had already become sufficiently established to be identified as both distinct and worthy of study. In his analysis, Long notes that liberation theologies can be distinguished by their central focus (e.g., women, the poor, etc.), but they share commonalities of approach. Thus, while they are different from one another in several ways, they bear analysis collectively and have common characteristics.⁴ They each present a theology distinct from orthodox Christianity and from each other, but generally similar characteristics of each allow for comparison between streams.

When Long wrote his analysis in 1982, ecotheology as a distinct Christian approach to environmental ethics was largely unknown. However, that changed with the broadening interest of ecofeminists, like Rosemary Radford Ruether.⁵ The crossover point for feminist liberation theology into ecological concerns is found in the search for a female deity to worship. In *Gaia & God*, Radford Ruether writes, "Ecofeminist theology and spirituality tended to assume that the 'Goddess' we need for ecological well-being is the reverse of the God we have in the Semitic

---

1. Olson, *The Journey of Modern Theology*, 507.
2. Gutiérrez, *A Theology of Liberation*.
3. Cone, *A Black Theology of Liberation*.
4. Long, *A Survey of Recent Christian Ethics*, 158–60.
5. Olson, *The Journey of Modern Theology*, 541. One definition of ecofeminism from one of its most vocal proponents reads, "Ecofeminism represents the union of the radical ecology movement, or what has been called 'deep ecology,' and feminism," Ruether, "Ecofeminism," 13. Ecofeminism is a form of feminist liberation theology that incorporates ecological concerns. There is sufficient overlap between the concerns and methodology between ecofeminism and ecotheology to warrant placement in the same category for the purpose of this paper. See Bernice Marie-Daly's work, which discusses the intersection of the ecofeminism and the ecology movement: Marie-Daly, *Ecofeminism*, 1–18. For more on the tie between ecofeminism and ecotheology in particular, see Conradie, *An Ecological Christian Anthropology*, 35–36.

monotheistic traditions; immanent rather than transcendent, female rather than male-identified, relational and interactive rather than dominating, pluriform and multicentered rather than uniform and monocentered."[6] Thus, for Radford Ruether, the importation of pagan notions of the deity is a part of the move to liberate both nature and women from patriarchy.

Olson treats Radford Ruether's ecofeminism under the heading of feminist liberation theology, perhaps to keep his categories simple.[7] However, his treatment may also be due to the perception that liberation theology has exhausted its energy. In 2010, John Frame commented, "Liberation theology was the dominant form of liberal theology from 1970 to about 2000. I think its attractive power is weakening somewhat."[8] This statement may be true of other forms of liberation theology, but not ecotheology. Contrary to Frame's observation, ecotheology as a form of liberation theology has only come into its own in the first decade of the third millennium.

Despite the rise of ecotheology as a distinct form of liberation theology, Olson is not alone in neglecting that fourth stream in scholarly work. Craig Nessan, a scholar who has written several volumes on liberation theology, neglects to document the rise of an ecological variant of the other three main liberation streams. It is not surprising that his 1989 volume *Orthopraxis or Heresy* does not discuss ecotheology, since the volume is documenting the North American response to Latin liberation theology and because of the ambiguity in the term "ecotheology" at the time it was written.[9] However, in a 2012 volume, *The Vitality of Liberation Theology*, which builds upon his older work, Nessan again ignores the stream.[10] This gap in such accomplished scholarship requires an explanation.

The rise of ecotheology as a form of liberation theology may have been obscured because it has its roots outside the church. Just as Radford Ruether draws inspiration from pagan sources to help make Christianity green, so ecotheology sometimes draws from indigenous sources and

---

6. Ruether, *Gaia & God*, 247.

7. Olson also neglects to deal with the significance of womanist theology, which is a blend between feminist and Black forms of liberation theology.

8. Sandlin and Frame, "Reflections of a Lifetime Theologian," 80.

9. Nessan, *Orthopraxis or Heresy*.

10. Nessan, *The Vitality of Liberation Theology*.

pagan conceptions to bring the idea of liberation of the earth to bear.[11] This dialectic synthesis, then, may have arisen in a sufficiently gradual fashion so that it has gone largely unnoticed in comparison to more traditional forms of liberation theology. A second possible reason for the relative obscurity of ecotheology may be that it arose outside of the traditional centers of theological innovation. Ecotheology as a distinct approach to Christian environmental ethics has its source with Ernst Conradie in South Africa and has only recently been adopted in the UK and Australia. It is now making more significant inroads into the theological discussion in the United States.

Another possibility for the relative obscurity of ecotheology as liberation theology may be the nature of the shift in theological thinking. In *The Journey of Modern Theology*, Olson describes the beginning of liberation theology as a paradigm shift, citing Thomas Kuhn's *The Structure of Scientific Revolutions*.[12] The introductory chapter of this project discusses Kuhn's paradigms in relation to the perspectives used herein. However, the method Kuhn describes for the spreading of paradigms is of greater significance here, because it may shed light on the undetected rise of ecotheology.

Kuhn notes, "When in the development of a natural science, an individual or group first produces a synthesis able to attract most of the next generation's practitioners, the older schools gradually disappear."[13] Kuhn's theory relates to natural sciences, but as Olson indicates, it appears there is a strong connection between the natural sciences and at least some schools of theology. Just as with new scientific paradigms in Kuhn's writing, ecotheology has seen "the formation of specialized journals, the foundation of specialists' societies, and the claim for a special place in the curriculum."[14] The existence of a sustained group at the Society of Biblical Literature's annual meeting on ecotheological hermeneutics[15] and

---

11. For example, McKay, "An Aboriginal Perspective on the Integrity of Creation," 213–17. This collected volume contains a variety of approaches to environmental ethics; it represents an early use of the word "ecotheology."

12. Olson, *The Journey of Modern Theology*, 504–06.

13. Kuhn, *The Structure of Scientific Revolutions*, 19.

14. Kuhn, *The Structure of Scientific Revolutions*, 19.

15. See Habel's personal explanation of his role in the rise of ecotheological hermeneutics: Habel, "The Origins and Challenges of an Ecojustice Hermeneutic," 141–59. He notes three dates of SBL consultations on ecotheological hermeneutics (2004–06); however, that consultation continued as late as 2015. Habel, "Introduction," 1.

the recent attempt by three ecotheologians associated with George Fox Evangelical Seminary to bring ecotheology into evangelical circles support this trend.[16] Time will be the ultimate judge regarding the reason for the subtle rise of ecotheology, but Kuhn's presentation of a paradigm shift has explanatory power in the near term. In fact, the etymology of the term *ecotheology* may provide evidence that supports a Kuhnian model.

Given the clumsy nature of other options like "Christian environmentalism" or "theological perspectives for environmental ethics," which appear throughout this project, it makes sense to create a shorthand term to refer to theology about a specific topic like the environment. This has led some to coin terms such as "theoecology" or "ecotheology."[17] Early in the use of the term, "ecotheology" simply meant theology done regarding the subject of ecology rather than implying a particular theological method. That meaning, however, has shifted over time.

Robert Fowler's 1995 survey of Protestant environmentalism lumps a wide variety of theological approaches to Christian environmentalism under the hyphenated term "eco-theology."[18] Similarly, in his 1996 book, *The Environment and Christian Ethics*, Michael Northcott uses the term for a broad range of theologies that all dealt with the environment. These ecotheologies include work done by Teilhard de Chardin, Francis Schaeffer, Jürgen Moltmann, Rosemary Radford Ruether, and others.[19] In contrast, Northcott's more recent volumes deal with environmental ethics from a theological perspective, but they largely refrain from using the term "ecotheology" to describe the effort.[20] Use of the term to refer to theology about the environment continues in social sciences, as in Katharine Wilkinson's volume, *Between God and Green*, where she refers to evangelicals doing "ecotheology" using traditional theological methods.[21] However, Wilkinson is doing history and not a theological analysis, so she may not be aware of the growing methodological significance of the term.

Northcott's avoidance of using the term "ecotheology" in his more recent volumes is likely due to the growing, though not unanimous, use

16. Brunner, Butler, and Swoboda, *Introducing Evangelical Ecotheology*.
17. George, "Theoecology–Definition."
18. Fowler, *The Greening of Protestant Thought*, 91–107.
19. See especially Northcott, "The Flowering of Ecotheology," 124–63.
20. For example, see Northcott and Scott, eds., *Systematic Theology and Climate Change*; Van Houtan and Northcott, eds., *Diversity and Dominion*.
21. Wilkinson, *Between God and Green*, 15–20.

of the word for methodology rather than content. There is still ambiguity in the use of the term, which has been aided by a lack of analytic work done on streams of environmental ethics within Christian theology.[22] It is impossible at this point to determine whether the use of the term "ecotheology" will come to refer solely to theological methodology in all fields. However, recent uses point in that direction. In the interim, recent scholarship has used uncertainty regarding the meaning of the term to attempt to draw more conservative Christians into a liberation approach to environmental ethics.

In their 2014 book, *Introducing Evangelical Ecotheology: Foundations in Scripture, Theology, History and Praxis*, Daniel Brunner, Jennifer Butler, and A. J. Swoboda attempt to present an ecotheological approach to environmental ethics as genuinely evangelical. The trio define themselves as evangelical based on the so-called Bebbington quadrilateral.[23] A closer examination of Bebbington's categories and ecotheology reveals a significant distance between the two, particularly in the category of biblicism. The distance between a robust biblicism and *Introducing an Evangelical Ecotheology* begins with the basic language of their theology. Consistent with the conventions of liberation theology, the authors eschew gendered language when referencing God.[24] The authors' choice of terms implicitly critiques the text of the Bible, which uses masculine pronouns to describe God. More explicitly, the authors admit that instead of taking the Bible as it presents itself, they are attempting to "embrace the Bible for its revelation of salvation and justice" while they "resist applying it through a model that views interpretation as timeless and transcultural."[25] In effect, they are subordinating Scripture to contemporary cultural standards, which is a hallmark of liberation theologies.

In general, *Introducing an Evangelical Ecotheology* presents an approach to theology that will seem palatable to an unsuspecting evangelical reader. Affirmations of "praxis"[26] and assertions of an "interconnection

---

22. At this time, the only critical essay on ecotheology appears to be Spencer, "Beyond Christian Environmentalism," 414–28. Portions of this chapter draw on the research for that article. My thanks go to Brian Tabb, managing editor of *Themelios*, for permission to re-use that research.

23. Brunner, Butler, and Swoboda, *Introducing Evangelical Ecotheology*, 5.

24. Brunner, Butler, and Swoboda, *Introducing Evangelical Ecotheology*, 21.

25. Brunner, Butler, and Swoboda, *Introducing Evangelical Ecotheology*, 21.

26. Brunner, Butler, and Swoboda, *Introducing Evangelical Ecotheology*, 7, 14, 19.

between ecology and liberation theology"[27] occasionally surface. This book represents a more subtle presentation of liberation theology than most from the ecotheological tradition, but its content is consistent with the praxis methodology of liberation theology and incompatible with traditionally accepted forms of evangelical theology. The substance of the arguments is supported by academics like Rosemary Radford Ruether, Sallie McFague, and Leonardo Boff—scholars who are openly skeptical, if not hostile, to the content of Scripture and traditionally orthodox Christian theology. Though Brunner, Butler, and Swoboda present their skepticism in more muted tones, an observant reader will find closer ties to the theological method of Ernst Conradie than the methods of theologians within the orthodox tradition.

The relatively recent shift in the meaning of the word "ecotheology" requires careful reading of earlier texts on Christian approaches to environmental ethics. Although Northcott uses the term, he is not referring to a liberation theology for the environment. However, the growing consistency of the use of the term for a particular theological method matches the definition used in this project.

A GENERIC ECOTHEOLOGICAL ENVIRONMENTAL ETHICS

More than the other theological perspectives discussed in this project, ecotheology stands apart from denominational structures. In other words, while there are denominations that are characterized generally as liberal, evangelical, or fundamentalist, there are no denominations that can be described as being ecotheological. That is because ecotheology lacks rootedness in any particular tradition, and instead emerges from the relatively recent postmodern movement.[28] Among its basic assumptions are the potential for epistemic shifts, a bias toward underrepresented voices, and suspicion of tradition. It is, in fact, profoundly different methodologically than the other theological categories discussed in this project; because ecotheology relies on a methodology that eschews stable

27. Brunner, Butler, and Swoboda, *Introducing Evangelical Ecotheology*, 93.

28. Olson notes an impetus, if not full acceptance, of postmodernism as a defining characteristic of liberation theologies: Olson, *The Journey of Modern Theology*, 545–46. Barton admits to relying on a postmodern epistemology in his reshaping of eschatology: Barton, "New Testament Eschatology and the Ecological Crisis in Theological and Ecclesial Perspective," 274.

expressions of truth, this discussion of the general characteristics of ecotheology will focus more on the theological method behind ecotheology than it will for other theological categories.[29] Despite the dearth of stable content within the ecotheology movement, the four questions that form a theological perspective on environmental ethics still apply and are still helpful in diagnosing an overall theological perspective for environmental ethics. This will become especially clear when examining the theology of Ernst Conradie below, but it is apparent even when we consider generic, ecotheological responses to the four questions under the doctrinal headings revelation, creation, anthropology and eschatology.

Revelation

For ecotheologians, the source of authority for environmental ethics is more diffuse than for the other three theological categories. It is a form of praxis theology, which means that personal experience plays a significantly greater role as authoritative revelation than in traditional theologies. Because of the immediacy of revelation through experience, Scripture and tradition are largely lumped into one source of authority, which is subordinate to the individual's experience. Scripture and tradition are mainly useful for illustrating and providing vocabulary for contemporary theological expressions. All of these sources of theological authority, which are bounded into the category of revelation, are subjected to pre-understandings, which are much more dependent on contemporary culture than on traditional theological sources.

Ecotheology is a version of contextual theology that interprets Scripture and Christian tradition through the controlling paradigm of the environment.[30] As a form of contextual theology, ecotheology can be

---

29. As with any contemporary stream of thought, one might say there are "many ecotheologies" in acknowledgement of the variations. However, despite the differences between the theological perspective of various ecotheologians, there is sufficient commonality to make some general statements about the movement.

30. Horrell finds a source of his hermeneutics in feminist theology: Horrell, *The Bible and the Environment*, 13. See also the explicit acceptance of the overlap between a feminist approach to Scripture and an ecotheological one in Horrell, Hunt, and Southgate, eds., *Greening Paul*, 11–12. For an expression of ecotheology that linguistically links ecotheology and liberation theology, see the report given to the World Council of Churches in 1991: "Liberating Life," 273–90. A similar link can be seen in Jenkins, "North American Environmental Liberation Theologies," 273–78.

cataloged within the praxis model.³¹ Stephen Bevans notes in his seminal work, *Models of Contextual Theology*, the praxis model "start[s] with the need either to adapt the gospel message of revelation or to listen to the context."³² Theologies in the praxis model "take inspiration from neither classic texts nor classic behavior, but from present realities and future possibilities."³³ As a form of praxis-oriented theology, ecotheology is consciously framed as different from traditional forms of theology, which are often viewed as Western or European and are thus considered foreign to much of the world and, consistent with White's hypothesis, bent on domination of ecosystems.³⁴

Bevans lists several presuppositions of the praxis model.³⁵ The most critical presupposition is epistemological, which Bevans lauds as the main strength of the praxis model. Praxis theologians begin from an understanding that "the highest level of knowing is intelligent and responsible doing."³⁶ Clodovis Boff explores this presupposition in detail in his oft-cited book, *Theology and Praxis: Epistemological Foundations*. Boff describes the ideal methodology of praxis theology, where right action is evaluated as the ultimate criterion of truth.³⁷ He is critical of

---

31. The term "praxis" is not without definitional difficulty. Praxis theologies include both practical outworking and theoretical development, so they are not purely practical. There are, as C. Boff points out, necessary theoretical foundations. In a discussion of the concept of the methodology of praxis theologies, Nessan identifies several key elements in a praxis form of theorizing: (1) contrast orientation to the life and experience of a particular population; (2) the use of social sciences for theological analysis; (3) reflection upon a population's conditions based on Christian tradition; and (4) actionable proposals for changing the present reality. Nessan, *Orthopraxis or Heresy*, 56–61.

32. Bevans, *Models of Contextual Theology*, 70. Bevans argues that all liberation theologies are praxis theologies, but not all praxis theologies are liberation theologies. An example of a non-liberation praxis model is the ecclesiocentric theology of Stanley Hauerwas; see Healy, *Hauerwas*, 39–72. Another example of non-liberation praxis theology can be found in Willis Jenkins's global ethic. Jenkins, *The Future of Ethics*.

33. Bevans, *Models of Contextual Theology*, 70.

34. See Nessan, *Orthopraxis or Heresy*, 3. Despite some differences in theological starting points, one of the close points of connection between liberation theologies and ecotheology is the rejection of economic development as a hopeful solution for the oppressed. Ecotheology rejects this because development *is* the oppression. Liberation theologies reject development because they feel that it *enables* the oppression of humans. Cf. Nessan, *Orthopraxis or Heresy*, 29–39.

35. Bevans, *Models of Contextual Theology*, 77.

36. Bevans, *Models of Contextual Theology*, 73.

37. Boff, *Theology and Praxis*, 195. Emphasis original.

this approach, as he notes, "To posit praxis as a criterion of truth implies empiricism and leads to pragmatism. It conjures away not only the theoretical problem, but the ethical one as well, which consists in asking *which* praxis and *which* theory are being referred to when this thesis is advanced."[38] In other words, Boff recognizes relying upon experience and practice as a criteria of truth instead of claims of orthodoxy merely shifts the point of theorizing and empowerment; it does not eliminate the existence of theory or the potential for oppressive use of power. Praxis must not, then, be confused with a mere practical theology.

Another key presupposition of the praxis model is that revelation from God is not contained in a static canon; rather, God is continuing to work throughout history in new and surprising ways.[39] This makes revelation available to all people at all times in the same way. No longer is God's special revelation limited to male authors who lived millennia before. These presuppositions are foundational to praxis theologies and undergird the theological method which drives their theoretical conclusions.

Ecotheology as a theological movement is consistent with Bevans's description of the praxis model. It is a theology that includes right action as a necessary component in its epistemic foundation, like all streams of liberation theology. Despite this similarity, one of the main ways ecotheology can be differentiated from other liberation theologies is the starting point. According to Nessan, "The starting point of liberation theology [in general] is most definitely the human situation."[40] For ecotheology, though the method is similar, the starting point is the condition of the created order, which requires a different set of theological presuppositions. A representative articulation of theological presuppositions of ecotheology can be found in the Earth Bible team's six ecojustice principles:

1. The principle of intrinsic worth: The universe, earth, and all its components have intrinsic worth/value.

2. The principle of interconnectedness: Earth is a community of interconnected living things that are mutually dependent on each other for life and survival.

---

38. Boff, *Theology and Praxis*, 231. Emphasis original.
39. Bevans, *Models of Contextual Theology*, 75–76.
40. Nessan, *Orthopraxis or Heresy*, 13.

3. The principle of voice: Earth is a subject capable of raising its voice in celebration and against injustice.

4. The principle of purpose: The universe, Earth and all its components are part of a dynamic cosmic design within which each piece has a place in the overall goal of that design.

5. The principle of mutual custodianship: Earth is a balanced and diverse domain where responsible custodians can function as partners with, rather than rulers over, earth to sustain its balance and a diverse earth community.

6. The principle of resistance: Earth and its components not only suffer from human injustices but actively resist them in the struggle for justice.[41]

A *prima facie* consideration of these six principles raises concerns about the use of such presuppositions when approaching the interpretation of Scripture and Christian tradition.[42] The method for choosing the presuppositions seems to be a significant question for ecotheologians. Ernst Conradie, largely supporting the approach, writes,

> The Earth Bible team acknowledge this danger but argue that each interpreter approaches a text with a set of governing assumptions that often remain unarticulated and subconscious and that are therefore even more dangerous. The danger of reading into the text randomly may be avoided if the articulation of such ecojustice principles is done in conjunction with historical, literary, and cultural modes of analysis.[43]

Conradie rightly notes that the six ecojustice principles prevent random reading into the text. They certainly direct the interpreters toward

---

41. Habel, "Introduction," 2. Similarly, David Horrell describes six theological foundations for his ecotheological interpretation: 1. The goodness of all creation; 2. Humanity as part of the community of creation; 3. Interconnectedness in failure and flourishing; 4. The covenant with all creation; 5. Creation's calling to praise God; 6. Liberation and reconciliation for all things. Horrell, *The Bible and the Environment*, 129–36. Mary Coloe notes that both of these sets of principles can be reconciled to one another: Coloe, "Preface," vii–viii. Such sets of principles serve as necessary foundations for beginning these conversations within a postmodern epistemology, because otherwise there would be no common starting point on which to base a conversation.

42. Liberation theologies tend to accept Bultmann's critique of attempts to exegete without presupposition. See Bultmann, "Is Exegesis without Presuppositions Possible?," 242–48.

43. Conradie, "Towards an Ecological Biblical Hermeneutics," 128.

consistent readings of the text, which resonate with the "perspective of justice for the earth."[44] Whether that perspective is in line with the divine or human authorial intent, the traditional interpretations of the canon, or the theological underpinnings of historic Christianity is another question. Given the overall suspicion toward the text of Scripture among ecotheologians, it is evident that consistency with historic Christianity is not considered a necessary good.

Ecotheological hermeneutics is a method of interpreting the Bible that regards the text with suspicion.[45] As such, the reader is called to believe the intentions of the human authors are corrupted by their context; thus, the text of Scripture cannot have a meaning that can be directly applied by the contemporary reader.[46] Instead of experiencing revelation mediated through Scripture, Bevans argues that revelation, in praxis theologies like ecotheology, is "a personal and communal encounter with divine presence" or "the presence of God in history."[47] This leads to a hermeneutic that is much less tied to historical readings of the text of Scripture and more concerned with finding immediate, present applications.

For ecotheologians, there is no difference between salvation history and ordinary history.[48] All of history is evidence of God working in time and space. This flattening of God's work in time and space presents a significant point of departure between traditional theology and ecotheology. Like other liberation theologies, ecotheology has a tendency "to collapse all of history into the imperious Now; to forget the paradoxical and serendipitous character of historical change; to downplay the provisionality of our historical moment and the partiality of our historical perspective."[49] Ecotheology, because of its emphasis on the importance of the contemporary context, often fails to truly listen to the voices of

---

44. Habel, "The Earth Bible Project," 123.

45. Conradie, *Angling for Interpretation*, 4.

46. Conradie writes, "Authors, texts, and readers are not 'innocent' or neutral," Conradie, *Angling for Interpretation*, 104. That the reader is not neutral is not controversial. That the authors may have been biased or clouded is also not questionable for extrabiblical sources. However, the concept that the text of Scripture itself is corrupt, even in the *autographa*, brings into question a great deal of Christian tradition.

47. Bevans, *Models of Contextual Theology*, 75.

48. Neuhaus, "Liberation as Program and Promise," 94. Neuhaus observes the continuity in history within liberation theologies in general, but it is applicable in particular to ecotheology.

49. Neuhaus, "Liberation as Program and Promise," 97.

Christian tradition. Instead, the goal is to hear the voice of the oppressed created order.

In his outline of ecotheological hermeneutics, Habel recommends intentionally attempting to retrieve the voice of earth from each biblical text. This may take the form of reconstructing the text with earth as the narrator. Habel notes, "Such a reconstruction is, of course, not the original text, but it is a reading as valid as the numerous readings of scholars over the centuries."[50] Habel demonstrates this approach in his attempt to retrieve ecotheological meaning from Genesis 1:26–28—notably one of the more difficult passages of Scripture for ecotheologians to redeem. In that passage, the author clearly describes God giving instructions to the primal couple to subdue creation and rule over the other creatures.[51] Since the passage is so clear, Habel simply questions the bias of the author, describing the passage as anthropocentric. This passage, he argues, gives rise to the ethical acceptance of domination and subjugation of the created order.[52] He then goes on to rewrite the passage, describing the cultural mandate from the perspective of the earth, "This story claims that the god-image creatures belong to a superior ruling class or species, thereby demeaning their nonhuman kin and diminishing their value. Instead of respecting me as their home and life source, the god-image creatures claim a mandate to crush me like an enemy or a slave."[53] The message of Genesis 1:26–28 is so distorted in Habel's retelling it is not clear how it can be described as connected to the actual Christian Scriptures. As Conradie notes in support of the Earth Bible project, "The assumed sacred authority of the Bible must therefore be questioned."[54]

Despite concerns about the actual authority of the Bible in ecotheology, there are a number of scholars who continue to attempt to redeem Scripture for their purposes. Habel and others have invested great energy in the Earth Bible project, which applies a circular theological method that bounces between Scripture and contemporary context with varying

---

50. Habel, "Introduction," 5.

51. Coming from an ecotheological perspective, Antoinette Collins argues that Gen 1:28 is necessarily negative toward creation, which is why she calls her audience to reject its message and instead understand earth care through the perspective of Australian aborigines. Collins, "Subdue and Conquer," 19–32.

52. Habel, "Introduction," 5–7.

53. Habel, "Introduction," 8.

54. Conradie, "Towards an Ecological Biblical Hermeneutics," 85.

weight placed on each as a source of authority.[55] However, ecotheologians approach authority for theology differently than the other theological categories. For ecotheologians, the Wesleyan Quadrilateral becomes a triangle. Scripture and tradition are essentially the same source of authority with no less authority granted to Calvin's commentaries than Paul's epistles. Ecotheology claims to be performing the necessary task of critiquing theological tradition. There is an assumption that today's interpretations will be revised and that the work of theology is to constantly remake Christianity in light of present concerns.

Even for conservative theologians, uncritical acceptance of historical theological interpretations is not a worthy goal. However, most theologically conservative Christians tend to critique historical theologians without rejecting the core understandings of orthodox theology. At the same time, traditional theologians tend to believe that contemporary theology should be done in conversation with historical theology, which implies that historical theology be allowed to critique contemporary trends.[56] However, there is little place in the methodology of ecotheology for the historic to influence the contemporary culture. Despite this, Bevans argues the praxis model has "deep roots in theological tradition."[57] Nessan affirms this assessment, arguing that liberation theologies are "deeply rooted in the Christian tradition" because they use the Christian Scriptures as a source for theology and due to "constant references to past formulations of the Christian tradition in articulating its own position."[58] Still, for ecotheologians, theology is not "a generally applicable, finished product for all times and in all places, but an understanding of and wrestling with God's presence in very particular situations."[59] There is a direct and inseparable tie between the substance of ecotheology and the context in which it is developed, and that tie is closer than the connection to

---

55. Most liberation theologians accept the circularity of their theological method, though more recent proponents describe it as a spiral. Regarding liberation theology and the hermeneutical circle: Boff, *Theology and Praxis*, 135–39; Nessan, *Orthopraxis or Heresy*, 62–63.

56. This is one of the reasons C. S. Lewis recommends the reading of books from previous centuries, since each period has its own blind spots and points of brilliance. Lewis, "On the Reading of Old Books," 202–04.

57. Bevans, *Models of Contextual Theology*, 78.

58. Nessan, *Orthopraxis or Heresy*, 402–03.

59. Bevans, *Models of Contextual Theology*, 78.

traditional Christian formulations. In fact, resistance to tradition is often supported in ecotheology as a resistance to power.

The Marxist roots of liberation theologies lead proponents to insist that all traditions and structures are attempts to seize and exercise power. The substance of liberation theology is a rejection of power structures, and thus a main thrust of ecotheology is to subvert structures that oppress the earth. Using a postmodern approach to truth, ecotheology attempts to give a voice to the marginalized earth. However, such attempts neglect the irony that by subverting the existing power structure, they are attempting to create a new power structure. The oppressed becomes the oppressor. It seems that new power structures, with the liberated earth at their heart, should be undermined, perhaps by a retrieval of authentic and faithful readings of the text and orthodox tradition by oppressed conservatives in the future.

Within the framework of ecotheology, reason—the third leg of the Wesleyan Quadrilateral—receives a fair amount of authority as long as it is based upon an approved foundation. The need for an approved foundation is why the interpretive efforts of the Earth Bible project assume their six principles. As the contemporary embodiment of reason, science is assumed to be authoritative, and little defense is given of this assumption.[60] It is simply accepted as part of the contemporary milieu and truth to which Christianity must adapt.[61] In the case of Christopher Southgate, this leads him to unapologetically reject the truthfulness of the account of the fall of humans in his ecotheological account of theodicy.[62] In other

60. See for example, Horrell's explicit calls to redefine Christianity in light of contemporary science: Horrell, *The Bible and the Environment*, 137, 43. Olson observes that liberation theologians make no attempt to reconcile Christianity and natural science, preferring to leave that to others and arguing it is a distraction for contemporary Christianity. Olson, *The Journey of Modern Theology*, 546. Olson's statement is generally true, though there are exceptions, as where Denis Edwards dislikes the pessimistic predictions for the burnout of the universe. In this case, Edwards recommends holding the contradictory views of New Creation in Scripture and a dead universe in science in tension. Edwards, *Jesus the Wisdom of God*, 146–48. However, Edwards specifically states that Christianity should not "reject scientific predictions about the future in a fundamentalist isolation from contemporary culture," Edwards, *Jesus the Wisdom of God*, 146.

61. The acceptance of evolutionary theory *en toto* leads Christopher Southgate to wrestle with theodicy and simply to accept suffering as part of God's initial design for the world. Southgate, *The Groaning of Creation*, 1–17.

62. Southgate, *The Groaning of Creation*, 28–35. Other ecotheologians, like Horrell, also abandon the traditional Christian understanding of the fall, but they attempt

words, where contemporary scientific explanations appear to conflict with Scripture, the preferred explanation is generally the one put forth by science.[63] Southgate deems Scripture that he perceives to be in conflict with scientific consensus to be "theologically unfruitful."[64] Therefore, he is willing to subject some difficult texts of Scripture to the superior authority of science and attempt to redefine Christian doctrine in light of contemporary assumptions.

Despite the precedence of contemporary science in ecotheology, it does not bear the greatest weight of authority. Like other liberation movements, priority of authority is given to experience in ecotheology. Authoritative experience is especially acquired through right living. Orthopraxy is more important than orthodoxy, so the point is not to get the facts right but to get people to act properly. This has led to an emphasis on pragmatic aspects of faith in the ethics of Willis Jenkins; good theology is doing things that resonate with an accepted set of values rather than those that match a set of dogmatic commitments.[65] Jenkins downplays the distinctiveness of Christian doctrines to the extent that he anticipates authentic expressions of true Christian praxis outside of confessional accord with basic Christian doctrines.[66] More than simply arguing for common grace, Jenkins is arguing that those that do not know Christ may be right with God due to their actions. According to Jenkins, "Social responsibility is not an expression or outreach of the church, then; it is partaking in Christ."[67] True theology is found in right living.[68] Theological tradition, even basic orthodoxy, is secondary to approved behavior

---

to do so textually or theologically. Horrell, *The Bible and the Environment*, 37–42.

63. Another example of a seeming unquestioned preference given to science is where Denis Edwards unapologetically and without defense simply assumes a naturalistic account of evolution with a 13.7-billion-year-old universe. This puts him in the position of attempting, like Southgate, to explain the goodness of God in light of the necessary suffering in evolution. Edwards, "Creation Seen in Light of Christ," 7–15.

64. Southgate, *The Groaning of Creation*, 18.

65. Jenkins, *The Future of Ethics*, 158–62.

66. Jenkins, *The Future of Ethics*, 99.

67. Jenkins, *The Future of Ethics*, 102.

68. Nessan differentiates the idea of praxis-based theologies from academic theologies. According to Nessan, the purpose of praxis is to establish a true consciousness instead of a false one, maintain a continuing stream of theological reflection, and motivate the audience to sustained behavioral transformation. This is much different from attempting to get at truth, which is the stated goal of orthodox theology. Nessan, *Orthopraxis or Heresy*, 408–10.

that is consistent with the contextual presuppositions. For ecotheology, these presuppositions will reflect an environmentally friendly lifestyle that serves the good of creation.[69] Environmentally responsible activity results in good feelings, which provide the moral authority for future environmentally responsible activities. Thus experience comes to dominate the other sources of authority for ecotheology, as long as the experience correlates to the basic philosophical presuppositions of the movement.

Ecotheology explicitly intends to reformulate Christian doctrines along environmentalist lines.[70] Authority is primarily rooted in the contemporary Christian's experience and understanding of the world, rather than in the content and historical interpretation of Christian Scripture. By shifting the foundation of doctrines, ecotheologians claim to be improving Christianity and adapting it to meet the needs of the contemporary context. In their attempt to renew biblical interpretations and revise Christian doctrines, they bring into question whether ecotheology has so reinterpreted tradition as to weaken the links between the traditional sources of meaning which have provided continuity and community for Christians across the millennia. In this way, ecotheology has created a theology that is more closely related to the contemporary cultural context than to the historic content of traditional forms of Christianity.

Creation

Regarding the value of the created order, the Earth Bible team's ecojustice principle assumes that each component of the created order has *intrinsic value*. In the brief defense provided for this principle, they allow the possibility that God imbued the creation with value. However, they conclude the section with what seems to be their preferred reading: "Earth and the components of earth in Genesis 1 are valued as 'good' by God when God discovers them to be so, not because God pronounces them to be so. In Genesis 1, Earth is 'good' of itself; this is a reality that God discovers."[71] This formulation for the value of Earth brings God's nature as an omniscient being into question, which is the most significant flaw in the approach because it opens the door to process theology.[72] More germane

---

69. E.g, Van Wieren, *Restored to Earth*, 152–55.
70. Conradie, "Towards an Ecological Biblical Hermeneutics," 133.
71. Earth Bible Team, "Guiding Ecojustice Principles," 44.
72. In fact, the Earth Bible team implies that it does not matter if the existing

to this discussion, it ascribes value to the created order apart from God's goodness.

The definition of value argued by the Earth Bible team is by no means unanimously held among ecotheologians. Some ecotheologians hold to value in the created order dependent on God. However, the means to describe such value tend to result in pantheistic or panentheistic themes. For example, Denis Edwards uses the term "intrinsic value" to describe the goodness of creation, but argues that "all things have value in themselves because of their relationship with God."[73] This definition allows Edwards to develop an environmental ethics that is livable, because human interests can be valued against those of nature.[74] Edwards's modification of the Earth Bible team's definition of "intrinsic value" is an improvement in his theological perspective for environmental ethics.

At the same time, using the term "intrinsic" leads Edwards to argue for value in creation in panenthestic terms.[75] Thus, he claims that mountains, forests, and ecosystems "have intrinsic value in themselves because they are modes of divine presence and the expression of trinitarian fruitfulness."[76] Even while arguing for total dependence on God by creation, Edwards simultaneously holds to the autonomy of all of creation. He finds himself at a point of tension here and argues that both God's absolute authority and creation's absolute autonomy must be held as supporting concepts.[77] Edwards's commitment to reconcile the traditional understanding of dependence of creation with the contemporary notion of its autonomy and self-worth is more consistent with traditional Christianity than some other forms of ecotheology. It allows him to speak the language of ecotheology but hold to a more traditional environmental ethic. However, it fails to establish a consistent place for other theologians to begin. This sort of variegation among ecotheologians is why the Earth

---

universe was in any real sense created by God. This lack of concern for the origin (not to say the timeline) of creation is enabled by the assumption that creation can have value apart from God. Earth Bible Team, "Guiding Ecojustice Principles," 47.

73. Edwards, *Jesus the Wisdom of God*, 154.

74. Sarkar, *Biodiversity and Environmental Philosophy*, 57.

75. He writes, "Creation is the interior relationship between the Creator and each creature by which the creature is held in being. If God were not interiorly present to each creature enabling it to be, it would be nothing," Edwards, "Creation Seen in Light of Christ," 8. There is a fine line between the idea that God sustains all creation (cf. Col 1) and panentheism, which Edwards appears to cross.

76. Edwards, *Jesus the Wisdom of God*, 162.

77. Edwards, "Creation Seen in Light of Christ," 9.

Bible team find it necessary to define the value of creation as intrinsic as a presupposition of their project.

A strength of intrinsic valuation of the created order within ecotheology is the rejection of a dualistic vision of the universe. There are some streams of Christianity, particularly popular Christianity, which have a view of the created order more consistent with a neo-Platonic perspective than a biblical perspective.[78] In response to this, ecotheologians argue toward a unification of all things. In some cases this results in a diminished or absent view of heaven. In Radford Ruether's work, creation maintains a spiritual nature, which appears to have more resonance with a pantheistic description of reality than orthodox Christianity. For example, in her vision of the afterlife, the disembodied souls of the dead rejoin a cosmic energy. In Radford Ruether's accounting, this destiny is common for all people apart from conversion, obedience, or any of the usual religious markers of faith.[79] Pantheism and panentheism are constant temptations to ecotheologians, as they erase distinctions between physical and spiritual conditions.[80] In large part, the departure from an orthodox understanding of the creator-creature distinction does not result from a direct pursuit of traditionally unacceptable views, but rather a weakening hold on those traditional views caused by an emphasis on action instead of doctrine.

The rejection of dualism of any form—even a hierarchical dualism that differentiates without divorce—is embedded in most ecotheologies. That perspective is enabled by an assumption of the autonomous, intrinsic value of the created order. Given this understanding of the value of the created order, an anthropology that is distinct from traditional Christian formulations is a necessary result.

Anthropology

One of the main differences between ecotheology and other streams of liberation theology is the diminution of anthropology, nearly to the point of extinction. While other liberation theologies typically begin with the

78. See Ruether's discussion of this: Ruether, "Religious Ecofeminism," 363–67.

79. Ruether, *Introducing Redemption in Christian Feminism*, 119–20. Note that here Ruether crosses from the value of creation into anthropological terms necessarily because she largely denies any differentiation between human and non-human creation.

80. Van Wieren, *Restored to Earth*, 78–80.

oppressed human at the center, ecotheology begins with the scarred earth as the central voice of concern. The distinction between human and non-human is largely erased by a rejection of hierarchy and dualism. Anthropology is largely ignored, except to critique concern for the uniqueness of humans within creation for being anthropocentric. The most significant treatment on anthropology from an ecotheological perspective is an attempt to minimize the doctrine and emphasize the human continuity with the created order.[81] Thus, fully half of the ecojustice principles of the Earth Bible team are dedicated to deconstructing traditional Christian theological anthropology.

Ecojustice principles two, five, and six deal most closely with the role of humans in the created order. Principle two highlights the idea of interconnectedness because, "traditionally, Western thought has assumed that male humans are creatures of a different order than other forms of life."[82] Both alleged sexism and anthropocentrism can be combatted with the principle of interconnectedness. Differentiation of humans from the rest of creation is diminished when the hierarchy, which is intuitively apparent in nature, is qualified by interconnectedness. In other words, even if humans are much more advanced in the ability to reason and interact than other animals, humans still must have other creatures to survive. This truism is used to flatten all of creation into one interwoven mass because, by this logic, the earthworm is equal in significance to the human. Raising the value of the created order to the level of intrinsic in effect merely reduces the value of humanity.

In principle five, the notion of mutual custodianship counters the anthropocentric concepts of human dominion or stewardship that are reflected in Genesis 1:26–28.[83] Mutual custodianship is an overt attempt to incorporate pagan ideas of kinship with the earth to minimize the sense of human responsibility within the created order.[84] The role of humans is now to avoid changing anything or negatively impacting other species.[85] This seems good on one hand, but its inconsistency with traditional readings of Scripture reinforces the limited authority of Christian tradition in ecotheology.

---

81. Conradie, *An Ecological Christian Anthropology*.
82. Earth Bible Team, "Guiding Ecojustice Principles," 45.
83. Earth Bible Team, "Guiding Ecojustice Principles," 50–51.
84. Earth Bible Team, "Guiding Ecojustice Principles," 51.
85. Earth Bible Team, "Guiding Ecojustice Principles," 51.

The Earth Bible team's ecojustice principle six, on the other hand, seems to weaken principle five. In order to reconcile the two, the reader must simultaneously hold onto the paradoxical concepts that humans are merely part of creation with no more rights and responsibility than any other part but that they are unique within creation as having the ability to exert special gifts to harm or help the rest of the created order, unlike other parts of creation. In light of this, humans—as just another part of earth—should join in resistance to ecological injustice. This results in tension between the impropriety of the human role as steward and the idea that humans "need to recognize that we have a moral obligation to help find a solution" to the ecological crisis.[86] Thus humans have a role, but that role is to actively pursue an undefined ecojustice and to passively allow earth to fulfill its purpose, which in ecotheology is generally described eschatalogically.[87]

Eschatology

The purpose of earth, according to ecojustice principle four of the Earth Bible project, is to exist according to its original design in perpetuity.[88] They assert, "The design is a magnificent green planet called Earth and the direction is to sustain life in all its diversity and beauty."[89] The team members are critical of some ecological accounts that focus efforts on maintaining ecological balance for the benefit of humans or other particular players within the interconnected web of Earth. They also call into question the concept of possible judgment and purging of creation. The Earth Bible team argues, "Within much of traditional Western Christianity, we learned to view the wonders of earth as but a foretaste of the glories to be experienced in heaven . . . Earth would eventually become waste, destroyed in God's cosmic incinerator."[90] In making this assertion, the Earth Bible team, with an allusion to Paul's words in Romans 8:18–23,

---

86. Earth Bible Team, "Guiding Ecojustice Principles," 53.

87. Horrell notes his own retention of such a form of anthropocentrism, but he fails to adequately explain how that perspective meshes with his overall theology. It is simply an assumed point of his theological perspective. Horrell, *The Bible and the Environment*, 137.

88. Earth Bible Team, "Guiding Ecojustice Principles," 49.

89. Earth Bible Team, "Guiding Ecojustice Principles," 48–49.

90. Earth Bible Team, "Guiding Ecojustice Principles," 49.

question the very content of Scripture.[91] Even clearer, they directly question whether 2 Peter 3:10 can be a valid understanding of the fate of the earth.[92] Thus, eschatology becomes an ethical mandate to return things to what they should have been all along. The difficulty remains in determining what the supposed original state of creation was so that it can be recaptured. Such an eschatological vision of repristination is made even more difficult because of tension between human responsibility to improve the status of the environment and human role as merely another part of the greater ecological web. As Horrell notes, despite the portraits in Scripture of universal reconciliation, "It is clear enough that reconciliation cannot mean an end to predation, nor indeed to competition (for scarce resources, living space, etc.), which is again intrinsic to the processes by which creatures co-exist and struggle to survive."[93] This apparent disregard for the clear content of Scripture (e.g., Isa 65:25) hearkens back to the question of the sources of authority for ecotheology.

Ecotheologians see Scripture pointing toward God's working to redeem all things. However, because of the concerns of anthropocentrism, ecotheology tends to strip eschatologies of their significance for the human portion of the created order. When ecotheologians address eschatology directly, it is often to refute the idea of a final conflagration of earth, leading to its utter destruction and a new creation. Beyond a rejection of a dispensational view of eschatology, ecotheologians tend to ignore aspects of judgment and discontinuity in eschatology, focusing on hope in cosmic restoration, though the nature of this restoration is presented in varying degrees of opacity. This may be attributed in part to their strong view of the continuity of history; not only are salvation history and ordinary history united, but there are no discontinuities within salvation history. To ecotheologians, just as the fall is a myth that represents the human experience of sin,[94] so the future judgments in the

91. Similarly, as part of the ecotheological hermeneutics project, Stephen Barton spends most of his essay attempting to show how traditional understandings of eschatology, which he allows are consistent with the New Testament, are unnecessarily pessimistic and need revision. Barton, "New Testament Eschatology and the Ecological Crisis in Theological and Ecclesial Perspective," 266–81.

92. Earth Bible Team, "Guiding Ecojustice Principles," 49.

93. Horrell, *The Bible and the Environment*, 143.

94. Ruether's idea of sin is telling in this regard: "My understanding of what sin is does not begin with the concept of alienation from God, a concept that strikes me as either meaningless or highly misleading to most people today," Ruether, *Introducing Redemption in Christian Feminism*, 70. Ruether goes on to argue that the primary

eschaton are representative of a type of redemption that is more progressive than cataclysmic.[95]

Radford Ruether deals with eschatology in her book, *Introducing Redemption in Christian Feminism*, but she rejects a traditional Christian understanding of the coming kingdom of God, where all things are made new. Instead, she posits a view of time that sees all existence in a cyclical manner, eternally changing. Thus, hope is not in renewal of creation and bodily resurrection. Rather, she asserts, "As we surrender our ego-clinging to personal immortality, we find ourselves upheld by the immortality of the wondrous whole, 'in whom we live and move and have our being.'"[96] Similarly, in liberation theologies, according to Lutheran theologian Carl Braaten:

> Eschatology is reduced to ethics. The kingdom of God arrives as a result of the ethical achievements of mankind. The gospel of the kingdom of God is removed to the future as a goal to be attained by the right kind of ethical activity. The gospel is not thought of as a present reality in history, already prior to human action, in the person and ministry of Jesus Christ.[97]

For ecotheologians, the kingdom is not now, but neither is it a distant reality to be brought into being only by dramatic intervention by God. The emphasis on this-worldly hope rather than judgment and renewal in the eschaton is likely a fulfillment of the evolutionary presuppositions of many ecotheologians, where the concept of a blemishless creation and subsequent original sin is discarded.[98] As Neuhaus points out, such a vision of the kingdom, with hope for salvation largely dependent on right living by humans, is illusory and worthy of critique.[99] Ecotheol-

---

concern is reconciliation of the horizontal, or creature-to-creature, relationship. Ruether, *Introducing Redemption in Christian Feminism*, 71–80.

95. Primavesi, *From Apocalypse to Genesis*, 73.

96. Ruether, *Introducing Redemption in Christian Feminism*, 120.

97. Braaten, "Gospel of Justification Sola Fide," 208. Also cited in Nessan, *Orthopraxis or Heresy*, 277.

98. Bouma-Prediger, *The Greening of Theology*, 150–54; Conradie, *An Ecological Christian Anthropology*, 30–31.

99. Neuhaus, "Liberation as Program and Promise," 92. For an extended critique of liberation theology in general on this point, see Nessan, *Orthopraxis or Heresy*, 270–83. Nessan works through the critiques of Neuhaus and Braaten in some detail.

ogy is, then, "in danger of transforming the gospel into a new synergistic scheme of salvation, a new form of revolutionary works righteousness."[100]

CONCLUSION

This section has presented an ecotheological perspective on environmental ethics by exploring four doctrinal questions. Other doctrinal questions are significant to the topic of environmental ethics, but a sufficient, basic understanding of the methodology beneath ecotheological perspectives can be gathered by considering these four questions. Of the four theological perspectives on environmental ethics, ecotheology is, by far, the most diverse because it has its roots in postmodernism. In general, a postmodern epistemology undergirds the idea of revelation. Thus, for most ecotheologians, revelation is immediate, direct, and comes from many sources besides the canon of Scripture. Creation tends to be valued in intrinsic terms, as if it has value in and of itself apart from God. Sometimes it is seen to have value that God discovered and not that he imparted. The role of humans in the created order, while not nil, is mainly to maintain the status quo or attempt to return to an earlier, more pristine state. Eschatology is primarily designed to motivate action, but does not refer to literal future events or even to a future state that is significantly different than the present. In broad terms the theological perspective of ecotheologians can be described in this manner. Given the diverse nature of ecotheology as a function of its very method, a more thorough examination of the theological perspective of Ernst Conradie will help to illustrate a common ecotheological response to the four questions that frame this project. Conradie's theology will be explored more deeply in the next chapter.

---

100. Nessan, *Orthopraxis or Heresy*, 277. Nessan is here summarizing the critiques offered by others of liberation theologians.

CHAPTER FOUR

THE ECOTHEOLOGY OF ERNST CONRADIE

ERNST CONRADIE IS, ARGUABLY, the father of the liberation perspective for environmental ethics.[1] He appears to have been the first to write about the methodology of ecotheology as liberation theology. Conradie serves as senior professor in the Department of Religion and Theology at the University of the Western Cape. He also edits the journal *Scriptura*.[2] With over one hundred and fifty refereed articles and essays in edited volumes—many on ecotheology—Conradie serves as a well-published source to represent the liberation approach to Christian environmental ethics.[3]

Because of the changing context in which ecotheologians write, it is more important to discuss the methodology at work than particular factual claims. The propositional content of theology will change based

---

1. Horrell describes Conradie's work as foundational to his project. David Horrell, "Introduction," 10.

2. "Faculty Biography of Ernst Conradie."

3. For example: Conradie and Jonker, *Angling for Interpretation*; Conradie, "A Preface on Empirical Biblical Hermeneutics"; Conradie, "Mission as Evangelism and as Development?"; Conradie, "Healing in Soteriological Perspective"; Conradie, "Climate Change and the Church"; Conradie, "Creation at the Heart of Mission?"; Conradie, "Justice, Peace and Care for Creation"; Conradie, "Climate Change as a Multi-Layered Crisis for Humanity"; Conradie, *Creation and Salvation*; Conradie, "Creation and Salvation"; Conradie and Jenkins, "Editors' Introduction: Ecology and Christian Soteriology."

on the available data.[4] The approach to this discussion will be somewhat different than in other chapters of this project. Given Conradie's significant, pioneering work in ecotheological hermeneutics, the discussion of revelation is couched in terms of hermeneutics instead of primarily in terms of the balance between science and Scripture. Just as in the generic ecotheological perspective, Conradie's sources of authority for theology conflate Christian tradition and Scripture. Science is generally assumed to be correct, and historical Christian theology must yield to its authority without question. Experience for Conradie is paramount. The foundational question of revelation and authority sets the stage for revisionist responses to the other three doctrinal questions.

After a discussion of Conradie's doctrine of revelation, this section probes Conradie's doctrines of creation, anthropology, and eschatology to establish Conradie's understanding of the value of the created order, the role of humans in the created order, and the end purpose for the created order. It is clear by examining Conradie's presentations of these doctrines how distant from traditional orthodoxy ecotheology tends to be in its present contextualized form.

REVELATION

Conradie admittedly mistrusts the Bible as a source of revelation.[5] He claims that the Bible is a dangerous book and that Scripture itself is ideologically distorted: it is not merely the interpretation of Scripture that is skewed, but Scripture itself.[6] Conradie uses this to explain why ideas such as specific gender roles and slavery, for example, are tolerated and sometimes seem to be extolled in the Bible. With regard to ecological theology, Conradie's chief concern is an anthropocentric bias.[7] He writes, "Beneath

---

4. In Conradie's description of ecotheological biblical interpretation he notes one of seven criteria for identifying the adequacy of an interpretation is its conditioning within "the contemporary context (societal challenges, changing circumstances)." Conradie, "What on Earth Is an Ecological Hermeneutics?," 299.

5. Conradie specifically affirms the notion that the biblical canon should be regarded with suspicion because "the texts are human products and therefore fallible. The aberrations and innovations which may be identified in the theological trajectories within the biblical texts are not necessarily normative for contemporary theological reflections. The biblical canon does not function as a fixated criterion for truth." Conradie, "On the Theological Extrapolation of Biblical Trajectories," 903.

6. Conradie, *Angling for Interpretation*, 29–31.

7. For example: Conradie, "The Road Towards an Ecological Biblical and

the tranquil surface [of the text], a power struggle is raging."[8] Conradie, therefore, proposes that the reader use a hermeneutics of suspicion when approaching the text.[9] He claims that his method "will enable you to see when something has gone wrong in biblical interpretation and where it has gone wrong."[10] However, the question of which text the reader should interpret remains open.

According to Conradie, contemporary readers should not trust the text of Scripture because of the absence of the *autographa*.[11] He argues that because there is so much debate about so many texts, the actual content of Scripture is so doubtful that it should not be trusted.[12] What readers can glean from Scripture are the attitudes of the authors, not the specific content of divine revelation. He cites the example of the creation account where, according to Conradie, "The confession of Israel's faith in Yahweh as the Creator in Genesis 1 does not provide us with a scientific report on the origins of the universe."[13] Rather, it provides an explanation of the attitude of wonder that the early Israelites had toward Yahweh and creation. Based on this position, contemporary Christians, if properly informed, are both free and able to judge the content of Scripture. Contemporary Christians certainly do not have to believe the content of Scripture.

In Conradie's words, "Christians do not believe in the Bible itself. They believe in the God of the Bible . . . Christians do not believe in the apostles whose witnesses may be found in the New Testament. They believe in Jesus Christ to whom the apostles bear witness."[14] He also criticizes

---

Theological Hermeneutics," 310.

8. Conradie also notes, "The 'hermeneutics of suspicion' wants to break this silence [of an undeclared power struggle] by explicitly speaking of things usually not mentioned. The emphasis on a power struggle implies that nobody can interpret from a neutral position. To interpret is, from the outset, to choose sides and to become engaged in the struggle." Conradie, *Angling for Interpretation*, 105. In another article, Conradie criticizes Scripture for the "anthropocentrism that has been present in the production of the biblical texts, that is evident in the surface structure of the text and has distorted the history of interpretation and contemporary reinterpretations of the Bible." Conradie, "Towards an Ecological Biblical Hermeneutics," 124.

9. Conradie, *Angling for Interpretation*, 21–32.
10. Conradie, *Angling for Interpretation*, 36.
11. Conradie, *Angling for Interpretation*, 62–63.
12. Conradie, *Angling for Interpretation*, 63–73.
13. Conradie, *Angling for Interpretation*, 58.
14. Conradie, *Angling for Interpretation*, 73.

"some fundamentalist Christians" for "attribut[ing] divine characteristics to the Bible. The Bible is regarded as equally trustworthy, authoritative, and inspirational compared to Godself."[15] Through this argument and others like it, Conradie dismisses the main stream of historical Christian understanding of Scripture.

Conradie's attitude toward the text of Scripture reduces the Bible to an occasional reference that can be considered authoritative in some sense only when the appropriate filters are applied to it. According to Conradie's description, the text of the Bible is not trustworthy; if it were, the substance of the text would be ideologically distorted anyway, which would thus prevent Scripture from being considered as a direct source of revelation. This suspicious attitude toward Scripture results in revisionist methodology being applied to the text itself in order to determine meaning, so that both biblical content and traditional interpretations are subject to revision or dismissal. This understanding of the authority of Scripture can be clearly seen in Conradie's well-developed ecotheological hermeneutics.

Ernst Conradie is a significant voice on hermeneutics for ecotheologians, since he was one of the first voices and has been a prolific advocate for a uniquely ecotheological hermeneutics. Studying Conradie's work is particularly helpful for understanding ecotheological hermeneutics because Conradie provides an explicit explanation of his hermeneutic through his books on the subject, as well as several other very pointed articles on hermeneutics.[16] Conradie emphasizes the ecotheological context in all of his writings, but there is really little difference between his methodology and that in feminist, Latino/a, or Black liberation theologies.[17]

Conradie's hermeneutic method consists of three steps:[18] first, attempting to articulate one's pre-understanding; second, beginning the ongoing process of critical testing; third, continually renewing earlier

---

15. Conradie, *Angling for Interpretation*, 73.

16. For example, Conradie, "What Are Interpretive Strategies?"; Conradie, "Towards an Ecological Biblical Hermeneutics"; Conradie, "What on Earth Is an Ecological Hermeneutics?"

17. Conradie, "Towards an Ecological Biblical Hermeneutics," 126; Conradie, "The Road Towards an Ecological Biblical and Theological Hermeneutics," 309.

18. Conradie presents a generic formula for interpretation, but himself wrote, "that interpretation cannot be reduced to a recipe, strategy or clear-cut method. It remains a form of art, a skill that can only be acquired and appreciated through experience." Conradie, "What Are Interpretive Strategies?," 440.

interpretations.[19] This model for what Conradie calls "adequate interpretation" is an explanation of a hermeneutic of suspicion. While a continual suspicion regarding the content and nature of revelation seems unsettling, Conradie describes it as similar to the institution of friendship. He writes, "Friendship is characterised by a relationship of mutual trust. Genuine friendship nevertheless requires both critique and suspicion. Likewise, a hermeneutics of trust requires critique and suspicion."[20] This tension between trust and suspicion is not discussed in more depth by Conradie, despite the apparent contradiction between continual suspicion and mutual trust.

Conradie's mistrust of Scripture and continual suspicion toward interpretation of Scripture leads to the question: What is Conradie's purpose in Bible study? Conradie seems to answer this question directly when he writes, "We discover the meaning of the Bible for today."[21] This seems to be a goal held in common with most expositors of Scripture; however, Conradie complicates the argument by writing a few pages later, "We cannot appropriate the instructions in the Bible directly for today."[22] Conradie teaches that meaning can be determined from Scripture, but that some indirect process must be followed to gain the material upon which his hermeneutical method can be applied. He suggests three interpretive strategies:[23] (1) find morals in the Bible; (2) find examples in the Bible that can be followed today; (3) find promises in Scripture that are being fulfilled anew today.[24] What Conradie seems to be presenting is a form of proof-texting (something he is highly critical of), performed

19. Conradie, *Angling for Interpretation*, 107. Cf. Conradie, "Interpreting the Bible Amidst Ecological Degradation," 200–01.

20. Conradie, *Angling for Interpretation*, 110.

21. Conradie, *Angling for Interpretation*, 88.

22. Conradie, *Angling for Interpretation*, 90. See also Conradie's comment that the actual product of the interpretive process is less important than "allow[ing] for a process of thorough critical testing in which as many critical questions as possible may be raised," Conradie, "What Are Interpretive Strategies?," 440. In other words, Conradie is more interested in the unsettling of previously-held theological interpretation than in determining appropriate theological interpretation.

23. Conradie differentiates the ideas of interpretative strategies and exegetical methods. He provides this definition of interpretative strategies: "Interpretative strategies refer to the ways in which readers establish a link between some aspects in the Biblical text and some aspect of the world in which they live." Conradie, "What Are Interpretive Strategies?," 429–30.

24. Conradie, *Angling for Interpretation*, 90. Also contained in Conradie, "What Are Interpretive Strategies?," 430–31.

by finding meaning in Scripture through acceptance of certain texts that support the interpreter's initial presuppositions. Conradie appears to commit the very sin of which he accuses others—namely, creating a breeding ground for ideologies as a means of gaining power, a condition that he claims results in distortions of the Bible.[25] The end goal of Conradie's hermeneutic methodology is to generate a "new, creative interpretation"[26] to replace the various contemporary and historical interpretations that Conradie declares inadequate.[27] The content of that new interpretation is determined by the desire of the interpreter using what Conradie calls "doctrinal keys."

Doctrinal keys, according to Conradie, are dependent on the theological pre-understanding of the reader's faith community. These keys are not based on Scripture, but rather on the faith community's previous attempts to resolve tensions between ancient text and contemporary context.[28] By Conradie's description, doctrinal keys appear to be concessions that the faith community has been forced to make in interpreting the text in order to prevent altering a cherished practice or a contemporary cultural belief.[29] Conradie argues that doctrinal keys are used not merely to help unlock the meaning of the ancient text for the contemporary reader, but more to make the text mean what the contemporary reader wants it to mean.[30] The difficulty in dependence on doctrinal keys is highlighted by Conradie, who notes, "The choice of doctrinal keys will necessarily (by

---

25. Conradie, *Angling for Interpretation*, 105.
26. Conradie, *Angling for Interpretation*, 118.
27. Conradie, *Angling for Interpretation*, 113.
28. Conradie, "What Are Interpretive Strategies?," 436.
29. At some times Conradie seems to approve of the use of doctrinal keys; at others he is very negative toward their use. It seems to depend on whether or not the doctrinal keys agree with his pre-understandings. He writes that doctrinal keys "regulate the interpretation of the Bible according to the so-called 'rule of faith' as determined by ecclesial authority," Conradie, *Angling for Interpretation*, 100. A prime example of Conradie's approval of the use of doctrinal keys is his review article of the Earth Bible Series: Conradie, "Towards an Ecological Biblical Hermeneutics: A Review Essay on the Earth Bible Project." He also affirms their use, especially with regard to contextual theologies in an article: Conradie, "The Road Towards an Ecological Biblical and Theological Hermeneutics." This is an amplification of his previous essay on the Earth Bible project. See also Conradie, "Biblical Hermeneutics of Liberation," 61–65; Conradie, "The Heuristic Key of 'Sustainable Community,'" 345–57. Conradie seeks to apply two major doctrinal keys simultaneously in an article on eschatology: Conradie, "Eschatology in South African Literature from the Struggle Period (1960–1994)."
30. Conradie, "What Are Interpretive Strategies?," 436.

definition) lead to a distortion of both text and context."[31] However, for this imposing difficulty, Conradie proposes no solution. For Conradie, there can be no permanent understanding of a text because the meaning is determined by the reader. The reader-centric nature of Conradie's hermeneutic is a central idea to remember when considering the application of the method described above.

Conradie's hermeneutical method illustrates the limited authority Scripture has in constructing an ethics of the environment for ecotheologians. Contemporary understandings of science and experience seem to be the dominant authorities in Conradie's theological perspective for environmental ethics, so that reshaping the content of Scripture to match a community's doctrinal keys is not simply authorized but encouraged. In much the same way, historical Christian doctrinal understandings are subject to revision and redescription to suit contemporary contexts.

In the ecotheological project, tradition is viewed with suspicion and referenced as a point of departure. Thus, as Conradie admits in his introduction to a volume discussing Abraham Kuyper's influence, he selected Kuyper as a conversation partner *despite* his disagreement with Kuyper's position on most issues. However, Conradie notes, "There were nevertheless some catch phrases in Kuyper's theology that were very appealing to me."[32] Conradie goes on to discuss how these catch phrases were appropriated, like mottoes, while the greater substance of Kuyper's theology was rejected. Much like the canon of Scripture, theological tradition is useful as a jumping-off point for new, creative interpretations, instead of providing a critique of current theological tendencies.

Conradie makes it clear that he is looking beyond the text for significance in Scripture when he writes, "The interpretation of the Bible cannot merely focus on the meaning of the texts themselves."[33] In fact, he states, "When one interprets the Bible, it is not only the text itself, but also the continuing historical influence of the text (also on ourselves) that is interpreted."[34] In other words, the meaning of text as written by the author is subordinate to the Christian community's reading of the text throughout history.[35] Also significantly, the meaning of the text as

---

31. Conradie, "What Are Interpretive Strategies?," 438.
32. Conradie, "Revisiting the Reception of Kuyper in South Africa," 24.
33. Conradie, *Angling for Interpretation*, 51.
34. Conradie, *Angling for Interpretation*, 83.
35. Conradie lists six things that must be considered in biblical interpretation: "1) The text itself; 2) The world-behind-the-text; 3) The history of interpretation

determined by the historic Christian community is subordinate to the contemporary community's determination of the meaning of the text. In order to provide a mandate for authenticity of revelation beyond the corrupted text, Conradie proposes some sort of ongoing revelation through and beyond Scripture. As he writes,

> God cannot be taken for granted or fixed into a set of abiding truths. In our ongoing relationship with God, God is free to act in new and surprising ways . . . . This requires an ongoing attempt to listen to the word of God found in the Scriptures. It is not an attempt to discover a set of eternal truths in Scripture but to hear the voice of the living God speaking anew to us today, in ever changing circumstances (through what God has done in the past).[36]

Conradie seeks what he considers a more palatable truth for the contemporary milieu, but this leaves him to reject any stable affirmations of truth.[37] In fact, Conradie claims that an attempt to teach any position, no matter how basic or directly drawn from the text, will, necessarily, result in "radical distortions in the interpretation of the Bible."[38] In this manner, the modern exegete can reject basic tenets of the faith as defined by the historical church while claiming to authentically represent historical, biblical Christianity, as Conradie does by including himself in the Reformed tradition. The diminution of the authority of the Bible and Christian tradition significantly influences all other doctrines in his theological perspective for environmental ethics, including his understanding of the value of creation.

---

in-front-of-the-text; 4) The spiral of appropriation and application; 5) The contemporary context(s) of the interpreter; 6) The possibility of ideological distortions from a world 'below' each of these aspects." Conradie, "Biblical Interpretation within the Context of Established Bible Study Groups," 444. While it is noteworthy that the first thing considered is the text, the five items considered have a strong potential to mask the text and reduce the text to a mere starting place for theology.

36. Conradie and Jonker, *Angling for Interpretation*, 89.

37. Conradie, *Angling for Interpretation*, 92. The possibility of a final (or even approximately so) interpretation is ruled out, in part because of the existence of a plurality of interpretation. See Conradie and Jonker, "Determining Relative Adequacy in Biblical Interpretation."

38. Conradie, *Angling for Interpretation*, 105. The unfortunate consequence of this position is that even the author's own position is undermined, since he is arguing for a particular method of reading the text that must be discarded as a radical distortion of the text.

CREATION

Conradie has done extensive research and writing on the doctrine of creation, including editing two volumes that attempt to unite the doctrines of creation and salvation. He notes,

> The need for an adequate theology of creation is typically taken for granted, given the familiarity of the theme in terms of the Christian confession. However, at times there has been a dangerous neglect of creation theology in order to focus, for example, on the existential and contextual relevance of the message of salvation, or on God's transforming mission in the world, or on secular processes of social transformation, or on the institutional needs of the church, or on a vague sense of spirituality.[39]

In his consideration of the doctrine of creation, Conradie notes there are many questions that can be answered under that heading. The question that is asked will depend greatly on the reason why the question is being asked. For example, from a cosmological position, the chief question is whether God made humans in his image or humans are making God in their image. Beginning with the relationship between science and theology, the nature of the evolutionary history of life is the chief question. Conradie's thesis about creation, then, is that "in God's eyes, that which is material, bodily, and earthly is precious to the Father, is worth dying for, and is being sanctified by the Spirit."[40] The most significant question for Conradie with regard to the doctrine of creation is why creation has value.

In his theology, Conradie rarely makes plain statements on any topic. This is, in part, due to the theological method, which consists of spirals of interpretation between text and culture, hoping to arrive at a useful meaning.[41] However, it seems that Conradie favors the Earth Bible team's understanding of the created order having intrinsic value.[42] He does not state this clearly, but instead Conradie offers a list of "models and images, each with biblical roots, with some considerable strengths but also some dangers," for describing creation.[43] Among these descriptions are creation as the fountain of life, God's work of art, God's gift to

---

39. Conradie, "What on Earth Did God Create?," 433.
40. Conradie, "What on Earth Did God Create?," 441.
41. Conradie, "What on Earth Is an Ecological Hermeneutics?," 298–301.
42. Conradie, "Towards an Ecological Biblical Hermeneutics," 128.
43. Conradie, "What on Earth Did God Create?," 447.

humans, God's body, and God's child.[44] Each of these offers a part of the explanation for value in the created order: for Conradie, creation appears to have value because of its intimate connection with God rather than on its own merit. The relational aspect of the value of creation is amplified by Conradie's understanding of the immanence of God within creation.

Conradie reveals that he is comfortable with the immanence of God in creation, even complimenting one author on finding common ground on this point with "neo-paganism, deep ecology, and constructive theology."[45] God's identification with creation helps establish the intrinsic value of creation for Conradie, much as it does for some of those that hold to a liberal theological perspective for environmental ethics. Although Conradie affirms the immanence of God within creation, he simultaneously maintains the importance of God's transcendence over creation.[46] For Conradie, however, transcendence takes on an interesting twist. The purpose behind Conradie's affirmation of the transcendence of God is not because of God's greatness in comparison to all created things, but rather because "the distinction of God and creation is important precisely for the sake of the integrity of creation."[47] He further explains that "the stress on God's otherness is crucial because it reminds us that the immanence of God can be understood in such a way that it would deprive creation of its freedom."[48]

The way he writes of freedom of creation illuminates Conradie's other statements on creation. For example, Conradie seems to argue that God sees creation as a subject, as God himself is a subject. In his statement of the origin of value in all of creation, including human creation, Conradie argues, "God does not love us because we are lovable; we become lovable precisely because God loves us."[49] God's relationship to the created order helps it to become lovable, and, it may be presumed, valuable. At the same time, the creator-creation relationship is almost ineffable. Conradie writes, "Suffice it to say that seeing the world through God's eyes as God's own beloved creation is very, very different from

---

44. Conradie, "What on Earth Did God Create?," 448–49.

45. Conradie, "How Can We Recognize God in the Singing River?," 149–50.

46. Conradie, "How Can We Recognize God in the Singing River?," 152–53.

47. Conradie, "How Can We Recognize God in the Singing River?," 153.

48. Conradie, "How Can We Recognize God in the Singing River?," 153n9. The referenced note is a cut-and-paste from a lengthier discussion in Conradie, "Towards an Agenda for Ecological Theology," 295–97.

49. Conradie, "What on Earth Is an Ecological Hermeneutics?," 445.

other ways of seeing it."[50] The emphasis here is that just as God is other, so is the created order other to God. This is an elevation of the created order to near personhood, which is consistent with Conradie's attempts to hear the voice of the created order. As Conradie develops the idea of the independence of creation, it forms a stronger basis for the idea of intrinsic value of creation than other approaches.

Conradie's concept of the value of creation is not a simple proposition. His ability to find value in God's creation due to God's declaration provides a better grounding for the concept of intrinsic value than is offered by deep ecology or other versions of the ecotheological movement. As discussed above, one major flaw in the Earth Bible team's principles is largely that they are merely assumed and not argued for. Conradie does provide an argument for his presuppositions, even using traditional theological terms. In the end, through careful argument, his concern for the freedom and independence of creation—its existence as a subject—leads him to similar conclusions to the Earth Bible team's about the independent value of creation.

ANTHROPOLOGY

Given the doctrines discussed so far, there should be little surprise that Conradie sets out to revise the Christian conception of anthropology. The basic categories of theological anthropology have been cast down by postmodern epistemology, according to Conradie.[51] In place of concerns about the constitution of the human person, the development of human technologies, or even human immortality, Conradie proposes more basic questions about how humans can know, how societies enforce power structures, and how a sense of community can be retrieved.[52] In other words, Conradie's anthropology has evolved significantly from theoretical concerns to more practical concerns. The great worldview questions of "Why?" are covered by new questions of "So what?" This is a contrast to traditional theologies, which reflects a dramatic revisioning of theological anthropology in Conradie's work and demonstrates how his approach tends to impoverish the ability to determine practical ethics by using an ecotheological perspective.

50. Conradie, "What on Earth Is an Ecological Hermeneutics?," 449.
51. Conradie, *An Ecological Christian Anthropology*, 3.
52. Conradie, *An Ecological Christian Anthropology*, 3–4.

PART TWO: ECOTHEOLOGY

For Conradie, anthropology is merely part of a theological schema in Christianity that he believes has been detrimental to environmentalism. He writes, "Christian piety has often inhibited an environmental ethos" including "a worldless notion of God's transcendence, a dualist anthropology, a personalist reduction of the cosmic scope of salvation and an escapist eschatology."[53] As a result, Conradie redefines the notion of humanity to eliminate the sense of alienation of humanity from nature, critique anthropocentrism of Christian anthropology, resist domination of non-human creation, and oppose any form of anthropological dualism.[54] Conradie offers this thesis statement for his anthropology: "One may summarise the thrust of this approach to ecological theology in terms of the thesis that human beings are 'at home on earth.'"[55] Thus, Conradie's anthropology is one of radical community between humanity and the non-human creation. In fact, his anthropology is opposed to such a distinction between aspects of creation. The solidarity of humanity and all of creation is drawn from Stoic philosophy, in which Conradie hopes to find a sense of place.[56] By establishing unity between all of creation, humans can thus be motivated to care for the *oikos* of God's creation. However, such an approach should be differentiated from the approach of deriving the doctrine of humanity from creation, which Conradie claims "builds an anthropology on the position of humanity before the fall (*status integritatis*)—a dispensation which no longer holds, if it ever did."[57] Thus, Conradie's anthropology attempts to describe things as they *are* instead of pointing toward the human condition *as it should be*.

Conradie identifies several key elements in his volume on ecological Christian anthropology. The first of these is that humans are creatures.[58] Conradie claims that traditional anthropologies fail to deal with the sense of alienation from nature, which has contributed to alienation between human beings and nature.[59] Instead, a properly developed ecotheo-

---

53. Conradie, *An Ecological Christian Anthropology*, 2.
54. Conradie, *An Ecological Christian Anthropology*, 5.
55. Conradie, *An Ecological Christian Anthropology*, 5. Original is italicized.
56. Conradie, *An Ecological Christian Anthropology*, 5–6.
57. Conradie, *An Ecological Christian Anthropology*, 10.
58. Although Conradie himself protests embedding anthropology within creation, he fails to recognize that those theologians that do not address creation and anthropology separately typically treat creation as a subsidiary doctrine to anthropology, not the other way around. For example, see Towns, *Theology for Today*, 555–86.
59. Conradie, *An Ecological Christian Anthropology*, 24–25.

logical anthropology should emphasize that "the human species forms part of the earth community. We are at home on earth."[60] As with the proponents of a liberal theological perspective for environmental ethics, Conradie relies on a cosmology that sees humans as latecomers to the created order.[61] Thus they are merely part of the created order and not a particularly important part. Also, humans may not be the end point of evolution, which should cause humans to be more humble and seek to find ways to live in community with nature. In other words, *homo sapiens sapiens* may simply be another step along an evolutionary chain to be replaced, like the Neanderthal in William Golding's *The Inheritors*.[62] Humans are just another part of creation, and should not see themselves as unique or distinct from the rest of creation. According to Conradie, "This community of all living species, including humans, is the greater reality and the greater value."[63] In ethical terms, then, the *summum bonum* for Conradie's ethics becomes interdependence because of the place of humans within the earth community.

A second aspect of Conradie's theological anthropology is a redefinition of the image of God. Instead of finding the value of humanity in the *imago Dei*, Conradie explores human worth through the notion of "human-in-relation-to-God."[64] Conradie argues that human dignity can be supported without strong claims for human uniqueness and without arguing for a special position of humans within creation.[65] His is not quite an argument for humans as mere animals, such as might be found in the discipline of evolutionary biology. On this point, however, Conradie appears to struggle with the apparent contradiction between the observable distinctions between humanity and the rest of creation and a desire to see humans as merely part of creation. He argues, "Humans are unique because we have been addressed by God in a particular way. One may immediately add that other species have been addressed by God too."[66] In this manner, Conradie sets humanity within creation but recognizes some distinction between humanity and creation.

60. Conradie, *An Ecological Christian Anthropology*, 26.
61. Conradie, *An Ecological Christian Anthropology*, 27.
62. Golding, *The Inheritors*.
63. Conradie, *An Ecological Christian Anthropology*, 39.
64. Conradie, *An Ecological Christian Anthropology*, 79.
65. Conradie, *An Ecological Christian Anthropology*, 80.
66. Conradie, *An Ecological Christian Anthropology*, 81.

Conradie emphasizes the technological developments of non-human species to support the idea that humans are merely a point on the evolutionary scale. Just as humans build houses, birds build nests. Just as humans use language and develop tools, some primates show forms of linguistic communication and have developed primitive tools.[67] From an evolutionary point of view, even among human species and their ancestors, language is only a recent development. But it was important because the development of language "allows for the evolution of religious rituals, myths, and experiences of transcendence."[68] Thus a part of the distinctiveness of humans is the ability to invent religions that enable humans to define themselves as distinct from all creation. In arguing the evolution of religion is part of the distinctiveness of humanity, Conradie undermines his own project. Environmental ethics done through a Christian lens becomes disingenuous if one assumes that all religious expression is merely invention. Such an approach invalidates the revealed nature of all religions, particularly Christianity, which may tend to explain Conradie's position on revelation discussed above. The deconstructive approach of Conradie's theological perspective threatens to undermine his argument for human participation in caring for creation.

In attempting to tiptoe along the fine line between uniqueness of humanity and a special position within creation, Conradie inevitably deals with the question of anthropocentrism rooted in being distinct from nature. Conradie rejects the notion that the *imago Dei* implies any sort of human transcendence over nature.[69] To do so, he argues, leads to the "impression that the universe was created specifically for our purposes and that the history of creation reaches its final goal with humankind. If that were true, the 14 billion years or so of God's creative love for creation is nothing more than a stage on which the drama of human salvation is worked out."[70] Conradie's understanding of the place of the human role in creation is one of democratization and community. Much like Aldo Leopold's land ethic, Conradie argues for "an affirmation of the *integrated* (if not equal) value of all forms of life (at the level of species) because of the integratedness of an ecosystem."[71] Such a perspective may be use-

---

67. Conradie, *An Ecological Christian Anthropology*, 99–103.
68. Conradie, *An Ecological Christian Anthropology*, 103.
69. Conradie, *An Ecological Christian Anthropology*, 94.
70. Conradie, *An Ecological Christian Anthropology*, 96.
71. Conradie, *An Ecological Christian Anthropology*, 127. Emphasis original.

ful in satisfying critics from outside the church, but arguing for such a continuity only raises questions regarding the human responsibility for environmental action. For Conradie, the answer seems to be that humans are to have as little impact as possible, leaving the earth in as close to pristine a state as can be. To do otherwise is to sin.[72]

Based on this approach and accepting the potential global devastation of climate destabilization as fact, Conradie finds such anthropogenic disturbance of nature's balance to be a prime evidence of sin. He contrasts this strongly to so-called natural environmental disasters like droughts, storms, etc., that cannot be traced to humans.[73] However, human responses to these disasters, often in an attempt to control or mitigate the damage, represent "one particular manifestation of structural evil, namely technological domination."[74] He does not deny the external benefits that such technological progress has had upon humans in general, especially with regard to improved quality of life and extended lifespans.[75] However, he is generally negative toward technological innovation because, in his calculus, the negatives of technology far outweigh the positives.[76] Since all of humanity is immersed in a technological world, this requires further exploration of Conradie's understanding of sin.

For Conradie, sin extends beyond "something which we do in the sense of specific wrongdoings" into "a self-destructive situation into which humans have collectively fallen."[77] In arguing this, Conradie embraces and even expands upon the notion of total depravity from his Reformed roots. However, the result is a blurring and widening of the category so it has very little value. As Conradie unpacks his doctrine of sin, he recategorizes sin as a fundamentally ecological phenomenon and denies the traditional understanding of original sin. Instead of seeing Adam's sin as a simple act of disobedience, the original sin is redefined as a denial of creatureliness.[78] At the surface this seems to be an acceptable change in terminology, since Adam and Eve fell for the temptation to be like the creator. However, given Conradie's denial of the substantive

---

72. Conradie, *An Ecological Christian Anthropology*, 183.
73. Conradie, *An Ecological Christian Anthropology*, 184–86.
74. Conradie, *An Ecological Christian Anthropology*, 187.
75. Conradie, *An Ecological Christian Anthropology*, 187.
76. Conradie, *An Ecological Christian Anthropology*, 188–90.
77. Conradie, *An Ecological Christian Anthropology*, 191.
78. Conradie, *An Ecological Christian Anthropology*, 193.

understanding of the *imago Dei*, this becomes problematic. Sin becomes an amorphous concept that is easily conflated with the notion of human finiteness, though Conradie denies their equivalence.[79] In the end, Conradie reduces sin to "the violation of relationships," which is, on the surface, consistent with traditional Christian orthodoxy, but it lacks sufficient content for repentance.

This lack of repentance is a significant gnomon in Conradie's notion of sin, which is apparent in his anthropology. Since human sin becomes (nearly) entirely corporate, the solution becomes corporate. Such a redefinition of sin diminishes the balanced biblical notion of corporate and individual responsibility, which allows a formalized, corporate activism to replace personal repentance. Thus, like other liberation theologies, ethics becomes mainly social and personal holiness becomes significantly underplayed. Conradie redefines repentance to be a change "in the human heart, in the collective psyche."[80] Thus, salvation in Conradie's ecotheology becomes of all creation and not of individuals. The place of damnation in Conradie's soteriology is left open, but that is a topic that is discussed in more detail under the heading of eschatology.

Since human sin is essentially the condition of being in distorted relationships, the ecological responsibility of humans is to live rightly with nature. Conradie accepts a stewardship motif for human responsibility to creation. He is careful to distance this from truncated versions of stewardship that deal primarily with the handling of money. Instead, he argues that "stewardship implies preservation and nurture."[81] Humans have the right to use natural resources for natural sustenance, but this must be done in a sustainable way. Conradie recognizes the common and oft-warranted criticisms of the theological concept of stewardship, but he argues for continued value in it.[82] He writes,

---

79. Conradie, *An Ecological Christian Anthropology*, 197–201.

80. Conradie, "Towards an Ecological Reformulation of the Christian Doctrine of Sin," 22.

81. Conradie, *An Ecological Christian Anthropology*, 209.

82. Among the weaknesses, Conradie sees that stewardship has potential to be used androcentrically, that it often elevates humans above other species, that it presumes authority and power and thus tends to be more useful for people in certain positions, and others. His critiques are valid, but in this case, it seems he is correct to affirm the value in the stewardship metaphor for human-creation interaction. See Conradie, *An Ecological Christian Anthropology*, 211–14.

> Perhaps this term may be translated and recontextualised in other ways which can build on the strengths of the metaphor of stewardship to emphasize human responsibility within God's household . . . Staying with God's household as the root metaphor, one may also bring metaphors such as housekeeping, homebuilding, housecleaning, home craft, home decoration, home-nursing, household tasks or simply "homework" into play.[83]

Despite bluster about anthropocentrism and potential abuse, Conradie lands on an environmental ethics that affirms a unique human role within the created order, though he uses a theological perspective distinct from other ethicists

Conradie begins to sound like more theologically orthodox Christians when he argues for the use of the stewardship motif and for some sort of unique role for humans within the created order. However, that image crumbles when Conradie unpacks the ethical principles he derives from his anthropology. For example, despite the unique role humans have within creation and even with the command to be fruitful and multiply (Gen 1:28), Conradie calls for limiting human population. He recognizes the powerful dilemma created by many approaches to environmental ethics: "To provide for the basic needs of the powerless, it seems that some further industrial development and growth is necessary—which will inevitably result in further depletion of non-renewable resources."[84] His solution is to reduce the birthrate as well as the rate of consumption, according to Conradie. He rightly notes that in his South African context, "the launching of birth control programs primarily aimed at under-privileged women have [sic] left a legacy of deep suspicion among the African population."[85] The challenge for Conradie is to simultaneously embrace God's love for abundant life while balancing the potential impact of humans on the earth.

Other than recognizing this delicate balance, there is little practical ethical content drawn from Conradie's anthropology. He points toward the need to have better ethics of gender and sexuality, but leaves these undefined. He also argues for the development of a careful ethics of food that recognizes the right to consume meat, but voluntarily limits it and seeks for just raising of the meat. Conradie also delves into the need

---

83. Conradie, *An Ecological Christian Anthropology*, 217.
84. Conradie, *An Ecological Christian Anthropology*, 233.
85. Conradie, *An Ecological Christian Anthropology*, 233n5.

for human work to enable holistic human flourishing. However, in the several pages of so-called practical ethics in his book on anthropology, Conradie reveals little that is specific and able to be put into practice. In the application section, particularly, he points toward the eschatological hope of humans, but there is little clear guidance given of the nature of that hope.[86] The difficulty in pinning down specifics for human action and justifiable boundaries of human stewardship illuminates the problem of a shifting theological foundation. In the end, a vague notion of future eschatological redemption is the strongest implication that can be drawn from Conradie's ecotheological anthropology.

ESCHATOLOGY

There is a strong connection between Conradie's anthropology and his eschatology. He argues, "We can only understand the place and vocation of humanity in the earth community if we have a sense of the destiny (*telos*) of creation and humanity."[87] Thus, the preceding discussion of anthropology largely depends on the present consideration of eschatology.

Eschatology is essential to Conradie's theological perspective on environmental ethics. As Conradie notes in the introduction to his theological anthropology, "Eschatology may indeed form the very key to an adequate ecological anthropology."[88] Thus, though Conradie attempts to show in his development of a doctrine of humanity that humans ought to strive to be at home on earth, he recognizes that the status of being "at home" can only be attained in some realized cosmic future. At the same time, despite his future focus, Conradie's eschatology is one that calls for humans to live as if the future is now. In this sense, his is an over-realized eschatology.[89] Although his language reflects the typical movements in Reformed theology of "creation, sin, providence, redemption, and consumption,"[90] Conradie's eschatology is much more immediate than other streams of the Reformed Protestantism, with which broad movement he associates himself.

---

86. Conradie, *An Ecological Christian Anthropology*, 232.

87. Conradie, "What Is Theological about Theological Anthropology?," 567.

88. Conradie, *An Ecological Christian Anthropology*, 13.

89. Conradie wrestles with the tension between already-not yet in Conradie, *An Ecological Christian Anthropology*, 223–30.

90. Conradie, *An Ecological Christian Anthropology*, 12–13.

Conradie's eschatological expressions are most clearly developed in his explanation of soteriology. His notion of soteriology is cosmic and not personal. Whereas most of the Reformed tradition reflects an understanding of the coming cosmic redemption, it has traditionally focused more heavily on personal redemption. This makes sense, at one level, because for the gospel to make sense cosmically, it must first be understood at a personal level. However, personal redemption is conspicuously absent from Conradie's soteriology and eschatology. This is, in part, due to his emphasis on sin as a general human condition and not a personal action.[91] He recognizes that there is individual sin, but emphasizes structural sin to a fault.[92] This emphasis on social over personal ethics is consistent with other liberation theologies.

There is more to Conradie's emphasis on cosmic eschatology than simply a focus on structural sin. It is rooted in the quest to redefine Christianity in eco-friendly terms. He writes,

> The track record of the history of Christianity in providing a form of hope that could empower an environmental praxis has not been too promising. Christian hope has often focused on the world to come, thus fostering and endorsing a sense of escapism and a lack of concern for this earth. Reinforced by apocalyptic images of the imminent destruction of the world in the biblical roots of Christianity, Christian hope has often been understood as final redemption *from* the earth and not *of* the earth itself.[93]

Thus Conradie's attempt to harness Christian tradition for the sake of the world requires a reconfiguring of the Christian tradition. He argues, "Biblical eschatology, with its unleashing of a dream of future perfection, is for many inimical to environmental concern."[94] And further, "It is the biblical injunction to transform the world that has inspired and legitimated ecological recklessness."[95] The ability to make such clear state-

---

91. Conradie, *An Ecological Christian Anthropology*, 191.

92. He argues, "The consequences of structural violence are typically more serious. The contemporary tendency is to view evil as the result of an accumulation, magnification, and institutionalisation of collective human wrongdoing, of insensitive decisions, unsustainable habits, and dangerous practices." Conradie, "The Salvation of the Earth from Anthropogenic Destruction," 117.

93. Conradie, "Towards an Agenda for Ecological Theology," 299. Emphasis original.

94. Conradie, "Towards an Agenda for Ecological Theology," 299.

95. Conradie, "Towards an Agenda for Ecological Theology," 299.

ments rejecting elements of the Christian tradition put Conradie and his eschatology in a difficult position.

Having rejected the Christian hope for eternal life as escapist, Conradie notes that many ecological theologies call for a focus on this life only. The problems with this are self-evident because, as Conradie himself notes, "A culture that assumes that there is nothing more than this life may easily degenerate into the caricatures of consumerism and hedonism or of cynicism and nihilism."[96] Conradie hopes to maintain within his eschatology a "vision of that which transcends this life" and a "hope for life beyond death," but it is unclear of what that hope rests on or consists of, given its distantiation from the account of Scripture.[97] As a result, in danger of being hoisted by his own petard, Conradie is forced to redefine the nature of salvation.

Conradie's vision of salvation is deliverance of all of creation from the impact of sin. In the emphasis on this theme, his vision is more consistent with the biblical presentation of eschatology than some of the merely personalistic eschatologies that sometimes emphasize a pseudo-gnostic disdain for the fallen creation. However, in attempting to correct the improper interpretation of Scripture in the Christian tradition, Conradie denies the message of Scripture and erases the foundation on which his position stands. Thus it is in the doctrine of eschatology that Conradie's flawed doctrine of revelation bears fruit and the extent of his deviation from traditional Christianity is revealed. At this time, Conradie has not resolved the questions that he has himself raised. The place of the individual and the nature of the future hope remain undefined.

What, then, is Conradie's understanding of the theological vision for the end state of the created order? It appears to be the redemption of all creation *from* human impact.[98] This is consistent with Conradie's concern for the impact of anthropogenic global warming.[99] Behind this is a vision of salvation as comprehensive well-being, which goes beyond daily physical needs, but Conradie is unclear as to what the nature and extent of that well-being is.[100] Instead of working to define clearly what the nature of eschatological hope is, Conradie seems to be encouraging Christians

---

96. Conradie, "Towards an Agenda for Ecological Theology," 300.

97. Conradie, "Towards an Agenda for Ecological Theology," 300.

98. Conradie, "The Salvation of the Earth from Anthropogenic Destruction," 113.

99. E.g., Conradie, "The Salvation of the Earth from Anthropogenic Destruction," 159–69.

100. Conradie, "The Salvation of the Earth from Anthropogenic Destruction," 115.

to find their own answers as long as such answers inspire an eco-friendly praxis. Thus his concern for orthopraxy instead of orthodoxy (a hallmark of liberation theologies, including ecotheology) is continued. The details of the future hope are sketchy (particularly when one rejects the validity of the limited definition offered by Scripture), but it has to be better than the present, so everyone should work together. This is why Conradie's theological perspective for environmental ethics is incapable of inspiring ethical impetus that speaks prophetically apart from the contemporary milieu.

CONCLUSION

Ernst Conradie's doctrine of revelation places human experience at the forefront of sources of authority, accompanied by contemporary interpretations of science. He shapes the language of his environmental ethics around the Christian tradition, in which he includes Scripture. Through this triage of authorities, he attempts to revise Christianity to liberate the oppressed earth from the sinful impact of humans.

Conradie ascribes value to creation intrinsically, though he is more moderate than other, more openly pantheistic ecotheologians. He simultaneously sees humans as merely part of creation and having a special role in creation. Humans have a special responsibility because God has uniquely gifted them and communicated with them. Eschatology serves the function of motivating action. It is thematic, but there is little expectation of actual personal salvation through the work of God. Instead, salvation is of all creation and is largely a metaphor for the need to free the earth of the effects of human sin.

Conradie's environmental ethics can be largely understood by answering the four questions regarding sources of authority, value of creation, role of humans in creation, and the future state of creation. These represent his theological perspective for environmental ethics, which provides a method for analysis and a point of contact for future engagement in dialogue.

PART THREE

A LIBERAL ENVIRONMENTAL ETHICS

CHAPTER FIVE

A Liberal Perspective for Environmental Ethics

This chapter examines a theologically liberal perspective for environmental ethics, which is best described as an attempt to balance the ethical claims of popularly accepted environmental ethics with the content of Scripture and church tradition. Christian environmental ethics as a sub-discipline originated with Joseph Sittler's work, rising to widespread attention with his 1961 address to the World Council of Churches. The earliest conversations on environmental ethics, and the largest volume of publication, are from theologically liberal sources. This makes the task of surveying a liberal theological perspective for environmental ethics a rich but somewhat daunting endeavor. This chapter discusses modernistic theological liberalism as it relates to environmental ethics. Then, in a general discussion of liberal environmentalism, it relates major themes in the liberal environmental ethics movement since its inception.

DEFINING LIBERALISM

As with the other descriptions of approaches to environmental ethics, the term "liberal" requires definition. First, it must be stated that the term is being used descriptively instead of pejoratively. There is an approach to Christian theology which can be categorized as liberal, both by those inside and outside the category, without intending a slight. In this project, the term is less significant than providing a category that reasonably groups similar approaches to theology. At times the term "revisionist" could be used to describe certain aspects of a liberal theology,[1] since a central aspect of a liberal theology is selectivity of the biblical witness, which often results in a revision of the voice of Scripture on some ethical topics.[2] "Modernist" may be another term, since the revision of Scripture and the shift from traditional understandings of theology have been significantly influenced by modernism, particularly in the use of culture and contemporary science as theological sources.[3]

In a discussion of common understandings of theological liberalism, Roger Olson notes, "Many people call any religious view they disagree with and think is somehow modern, as opposed to traditional, liberal theology."[4] That is decidedly not the definition of the term being used in this book. Instead, while recognizing the fuzzy boundaries of liberalism (just like every other theological perspective within this project), the word "liberal" is being used here to refer to a theology with content that maps in a particular range along a spectrum and which relies on a fairly consistent methodology that is distinct from the other three theological perspectives discussed within this volume.

Roger Olson provides a useful rubric for evaluating liberal theologies. Olson describes four major themes that characterize liberal theologians: (1) acknowledgment of modernity as a source for theology, (2) focus on the immanence of God, (3) moralization of dogma, and (4) universal salvation of all humanity. It is Olson's framework that will be applied in this chapter.[5]

Each of these four dimensions in Olson's rubric requires some consideration. The acknowledgement of modernity as a source for theology brings the so-called Wesleyan Quadrilateral into the conversation. In the case of liberalism, as Olson describes it, modernism is not merely a

---

1. See Santmire's description of his project, which is a revision of traditional Christianity for ecological purposes: Santmire, *Nature Reborn*, 1–15.

2. In what amounts to a popular-level hermeneutics text for liberals, Harvey Cox demonstrates this selectivity of the text and submission of the text to the authority of current understandings of science and ethics is an essential element of mature faith. This is most clearly evidenced in his instructions on reading Exodus. Cox, *How to Read the Bible*, 41–64.

3. Olson, *The Story of Christian Theology*, 550.

4. Olson, *The Journey of Modern Theology*, 125.

5. Olson, *The Story of Christian Theology*, 550.

source for theology along with Scripture, but also it becomes a source in competition with and sometimes supersedes the traditionally supreme role of Scripture within the Protestant tradition. In some cases, Protestant Christian environmentalists use modern science to "substantiate the claim of a current crisis, to provide the basic outlines of nature (of the biological and physical worlds), and to model the good society."[6] By defining both the problem and the solution according to science, this modernistic methodology usurps the traditional role of Scripture in theology.

At its base, this approach puts humanity in control of its own destiny and reduces the impact of appeals to authority for theology. There is a bias against authority within the modern liberal ethos, as sociologist Jonathan Haidt points out in *The Righteous Mind*.[7] Since the role of authority has been diminished in the contemporary liberal mindset, there is an obvious conflict for liberals who identify as Christian with the traditional authority of the Bible.[8] The resolution to the conflict is found by bringing contemporary authorities, which can be controlled and more readily debated, alongside the historically accepted centrality of the authority of Scripture. This allows liberal theology to remain connected to tradition, but with modernism as a control.

Of the time prior to the rise of theological liberalism, Olson writes, "Never before in Christian history had any theologian openly admitted that his or her culture rightly functioned as a norm for theology's content."[9] There is little doubt that culture has always influenced theology, but the rise of modern liberalism marks the beginning of embracing that reality. This helps explain the deep division between fundamentalists and liberals along theological lines. Modernism allowed both groups to recognize the significant influence of context on their theology. Fundamentalists responded largely by attempting to be as objective as possible, while liberals accepted the epistemic limitations on human objectivity and embraced subjectivity by affirming contemporary culture as a theological source. In part, for the liberal, modern culture becomes a source for theology because God is always near.

The emphasis on the immanence of God within liberalism has significant implications for environmentalism. Olson notes,

6. Fowler, *The Greening of Protestant Thought*, 5.
7. Haidt, *The Righteous Mind*, 345–46.
8. Gustafson, "The Place of Scripture in Christian Ethics," 164–65.
9. Olson, *The Journey of Modern Theology*, 127.

> Nearly all liberal Protestant thinkers of the late nineteenth and early twentieth centuries emphasized continuity between God and nature in a way that often smacked of pantheism or at least panentheism (mutuality between God and the world). At its best, liberal theology stopped short of reducing God to a universal World Spirit and affirmed God's personal nature. But always the imagery of humanity and nature as somehow extensions of God arose in the background if not the foreground of liberal thinking.[10]

Olson's criticism resonates with Lynn White's call to reshape Christianity as pantheistic. White, a historian, likely sensed the impact of God's immanence on certain forms of Christianity and was essentially urging liberal Christians to continue along their trajectory. This emphasis also helps to explain why environmental ethics as a Christian discipline has found in liberal theology an excellent agar.

Setting aside the danger of panentheism, this notion of immanence of God in nature has implications for the value of the created order. As the hierarchical distinction between creation and God dwindles, nature can be ascribed *intrinsic* value—that is, value in and of itself. As will be discussed in detail below, not all liberal theologians describe the value of creation as intrinsic, or, if they do, they may mean something more like *inherent* value. However, in general terms, there is openness within the liberal theological perspective for a closer connection between the created order and God than more conservative Christians typically allow, which Olson gives witness to when describing the relationship between liberalism and pantheism or panentheism.

The focus on immanence also often results in an over-realized eschatology. Combined with the notion of universal salvation, the immediacy of the *eschaton* implies that the experience of God is not contained within the boundaries of Christianity. Also, there tends to be a focus on all people at all times being able to equally receive revelation from God.[11] While this sounds very egalitarian in principle, in practice it tends to subject theological liberals to the tyranny of the present. It also makes discourse about distinctly Christian approaches to any ethical topic, including the environment, exceedingly difficult. In some cases, this tendency leads to the belief that the eschatological kingdom can be obtained in this life.

10. Olson, *The Story of Christian Theology*, 550.
11. Olson, *The Journey of Modern Theology*, 127.

Liberal theologians, then, sometimes critique the futuristic eschatological hope of those more conservative than themselves. Norman Wirzba writes, "For too long too many Christians have thought that the point of faith is to prepare people to enter a heavenly realm 'somewhere beyond the blue.'"[12] The hope of a present, fulfilled *eschaton* is seen especially in the Social Gospel movement, for whom Walter Rauschenbusch is, perhaps, the patron saint.[13] According to Harry Heubner, for Rauschenbusch, "in the coming of the Messiah the new reign has begun, and while not something we manage, we are invited, and indeed mandated to join in the movement."[14] The immediacy of eschatology is a theme within contemporary versions of liberal theology that allows some to see the kingdom of God being evidenced among those serving the environment who are not explicitly Christian and even explicitly not Christian.[15]

According to Olson the moralization of dogma is the third common theme for liberal theologians. He writes, "Under the influence of Kant, liberal Protestant thinkers insisted on reinterpreting all doctrines of Christianity in ethical and moral terms, and those that could not be so reinterpreted were neglected if not discarded entirely."[16] Olson's prime example of neglect is the doctrine of the Trinity.[17] There is evidence of attempts to correct this, even within liberal theological approaches to environmental ethics. For example, Timothy Gorringe both critiques Kant's assertion of the impracticality of the Trinity and attempts to present the Trinity in moral terms.[18] The focus of his moralization is to find a basis for combatting climate change through the doctrine of the Trinity. This is a worthy attempt at bringing the Trinity to bear on climate change, but Gorringe undermines his approach by presenting three other doctrines through his Trinitarian analysis: creation, revelation, and eschatology.[19] Gorringe presents a systematic approach to the doctrine of the Trinity that is explicitly moralized, but the theoretical discussion serves largely

---

12. Wirzba, *From Nature to Creation*, 1.
13. White, *The Changing Continuity of Christian Ethics*, 292–95.
14. Huebner, *An Introduction to Christian Ethics*, 321.
15. For example, see the claims of Jenkins, *The Future of Ethics*, 97–102.
16. Olson, *The Story of Christian Theology*, 550.
17. Olson, *The Story of Christian Theology*, 551.
18. Gorringe, "The Trinity," 15.
19. Gorringe, "The Trinity," 15–28.

as a platform for exploring more practical doctrines, such as those represented in a theological perspective for environmental ethics.

As a part of the emphasis on moralizing dogma, praxis is also emphasized within the liberal theological perspective.[20] Though certainly not to the degree that can be seen among ecotheologians, much of the effort in liberal forms of environmentalism is to inspire action. For "green Christians," Fowler writes, "the overarching assumption is always the urgency of the need for action, an unrelenting theme among [liberal] Protestant environmentalists." The uniting of faith and action is commendable. According to Fowler, "The downside, however, can be a neglect of reflective thought in the rush to action. In this instance, such a tendency is exaggerated, because so much of [liberal] Protestant environmental effort has as its source an overpowering sense that time is short and the crisis overwhelming."[21] The focus on environmentalism over evangelism in some forms of eco-friendly Christianity is enabled by a tendency toward universalism in salvation, which is part of the liberal tradition.[22]

The temptation of universal salvation in environmentalism is especially strong because of the emphasis on cosmic restoration. Steven Bouma-Prediger, himself not a liberal but a progressive evangelical, illustrates the slide toward universalism as he exegetes Col 1:20.[23] He argues, "Christ's work is as wide as the creation itself. It is nothing short of the restoration and consummation of all creation . . . Indeed, if Jesus did not die for white-tailed deer, redheaded woodpeckers, blue whales, and green Belizean rain forests, then he did not die for you and me. Jesus comes to save not just us but the whole world."[24] He goes on, "This salvation of all things, accomplished on the cross, is vindicated in the resurrection. The resurrection pertains not only to people; it embraces the earth."[25] The restoration of all things through the power of Christ quickly becomes

---

20. Indeed, Peter Kreeft notes that modernism reduces all religion to ethics. Christianity becomes *merely* a way of living and not a way of attaining the next. Kreeft, *Back to Virtue*, 32.

21. Fowler, *The Greening of Protestant Thought*, 142.

22. Langford, *The Tradition of Liberal Theology*, 50–51.

23. Though Bouma-Prediger is not himself a liberal, he was significantly influenced by Joseph Sittler, Juergen Moltmann, and Rosemary Ruether, as they were the subjects of his doctoral dissertation. Bouma-Prediger, *The Greening of Theology*.

24. Bouma-Prediger, *For the Beauty of the Earth*, 124.

25. Bouma-Prediger, *For the Beauty of the Earth*, 124.

the salvation of all creation. When insentient trees are being saved from the effects of sin, it becomes difficult to imagine the atonement not being universally applicable even to humans who actively sin.[26]

In most cases, liberal environmental ethicists simply do not address the possibility of human damnation in their writing. Universal salvation seems to be assumed.[27] As Sean McDonagh notes, "The theology of redemption in recent years no longer speaks of saving souls but attempts to recapture the all-embracing understanding of the world which is present in the Bible."[28] Like Bouma-Prediger, McDonagh uses Colossians 1:20 to claim universal redemption, which he appears to connect more readily to all humans apart from explicit faith in Christ.[29]

As Langford notes in his explanation of the liberal tradition of accepting salvation outside of explicit faith in Christ, "In addition to God's grace, the one absolutely essential thing required for salvation . . . is to *endeavor* to do the good, as one sees it. This is to respond to the word of God."[30] He goes on to argue, "All humans, whether of any other faith or none, who respond to God's word as it comes to them in their particular circumstances, whether before or after the time of Christ are in reality responding to the *Logos*, to the eternal Christ, whether or not they realize that this is so."[31] Salvation is thus a function of some effort to be righteous with good intention. Thus propagation of the gospel need not be a major emphasis of Christianity, so social justice issues can become the primary interest of those with a liberal theological perspective. The basic characteristics of liberalism, as Olson describes them, encourage a rapid,

---

26. This is not to say that Bouma-Prediger supports universalism, but to show how easy the drift becomes. It appears that his language in this case could have been tightened to ensure his meaning was not masked.

27. Where others speak in vague terms, leaving doubt about their understanding of personal salvation and damnation in light of cosmic redemption, Matthew Fox makes his position plain. He writes, "The idea of a private salvation is utterly obsolete. Only a Newtonian worldview of piecemealness could have spawned the popular heresy that salvation is an individualistic or private matter . . . Salvation must be universal in the sense of comprehensive, a healing of *all the cosmos' pain, or it is not salvation at all.*" Fox, The Coming of the Cosmic Christ, 151.

28. McDonagh, *To Care for the Earth*, 126.

29. McDonagh, *To Care for the Earth*, 127–28. McDonagh makes no claims about the scope of salvation of individual people explicitly; however, his concern for learning from other traditions without attempting to reach them with the message of Christ is telling. McDonagh, *To Care for the Earth*, 143–53.

30. Langford, *The Tradition of Liberal Theology*, 35. Emphasis original.

31. Langford, *The Tradition of Liberal Theology*, 35.

socially conforming response to many social justice issues, including environmentalism.

For the issue of environmental ethics in particular, Olson's four themes of liberal theology work themselves out in a contextual fashion: (1) acknowledge science and experience as primary sources for environmental ethics, (2) (a) focus on the need for human rather than divine intervention in nature and (b) see nature as valuable in itself, (3) apply doctrines thematically instead of particularly, and (4) prioritize the pursuit of environmental justice over personal evangelism. These are four expressions of the principles that define theological liberalism, according to Olson's outline. They form a part of the theological foundation that can be observed within the liberal theological perspective for environmental ethics.

A GENERIC LIBERAL ENVIRONMENTAL ETHICS

The literature on environmental ethics from a liberal Christian perspective is much more expansive than any of the other theological categories discussed in this project.[32] This is a result of a greater interest in theologically supporting environmental activism, which is consistent with greater political engagement with social causes that are often considered liberal. It should be clear, however, that while a liberal theological approach tends to engender an increased interest in environmental ethics, an interest in environmental ethics does not require such a liberal theological method. As my social media mentions show, there is a common tendency to associate concern with the environment with liberalism; it is generally true that more theological liberals than conservatives engage in environmental ethics, but such a correlation does not imply causation.[33]

An apology is necessary at this point, or at least a *caveat lector*. Even more so than in the fundamentalist and evangelical perspectives offered in later chapters of this volume, there is variation within the liberal theological stream. Thus, this section will speak in general terms in an

---

32. Robert Fowler made this observation in 1994. It is even more true today. Fowler, *The Greening of Protestant Thought*, 14, 51.

33. For example, Smith and Johnson note that claims of the liberalization of young evangelicals are often drawn from an increased engagement in environmental ethics, though the perceptions of young evangelicals are largely consistent with the older generations on other moral issues. Thus, their conclusion is that the trend is overstated. Smith and Johnson, "The Liberalization of Young Evangelicals," 358–59.

attempt to define what is roughly the center of the theological stream. The extended definition offered in the previous section helps to bound the category, and the representative sources used in this section have been chosen because in general they fall clearly within the categories or self-identify as liberal.[34] In any case, the label is here intended to describe a theological tendency rather than to diminish the validity of the theologians' contributions.

Despite the number and variety of liberal voices that have entered the discussion, there are general patterns that appear in introductory theological treatments of environmental ethics from within the liberal Christian tradition.[35] The volumes often begin in the introduction with a conversion narrative; the author has an experience of witnessing human-caused environmental degradation, which evokes an emotional response.[36] A case is made for the emergent nature of the degradation through accounts, statistics, and scientific reasoning.[37] The author argues that Christianity is, or has potential to be, positive for the environment.[38] Then follows an explanation of how that works out through new

---

34. Some of these decisions are more difficult than others. For example, in this chapter I have identified Michael Northcott as a theological liberal, although he was influenced by Francis Schaeffer earlier in his life, and his earlier writings seemed more consistent with European evangelicalism. However, his more recent work, particularly his co-authored introductory chapter in the volume *Systematic Theology and Climate Change*, indicates a firm stance that fits within the liberal category as it is defined here. In one case, Northcott and Scott write, "In this new context, religion, and especially religions like Christianity which are open to creative doctrinal reformulations in new contexts, offer cultural resources for enduring, and finding meaning in, the new world of the Anthropocene." Northcott and Scott, "Introduction," 5.

35. These characteristics are typical of introductory texts because they are designed as an apologetic for a form of Christianity for non-Christian environmentalists. They are also intended to draw interested Christians into the environmental conversation, while allowing them to maintain the connection with their Christian tradition. The particular order, with science discussed before Scripture, reflects a subversive sense of superiority for Science over Scripture, whatever else the author may say. See Fowler, *The Greening of Protestant Thought*, 5.

36. Kureethadam, *Creation in Crisis*, ix–xi; McDonagh, *To Care for the Earth*, 1–13; Nash, *Loving Nature*, 11–16.

37. Kureethadam, *Creation in Crisis*, 15–291; McDonagh, *To Care for the Earth*, 17–106; Nash, *Loving Nature*, 23–67; Northcott, *The Environment and Christian Ethics*, 1–85; Rasmussen, *Earth Community Earth Ethics*, 25–180.

38. Kureethadam, *Creation in Crisis*, 293–328; McDonagh, *To Care for the Earth*, 107–42; Nash, *Loving Nature*, 68–138; Northcott, *The Environment and Christian Ethics*, 124–63; Rasmussen, *Earth Community Earth Ethics*, 181–321.

theological expression and through activism.[39] There is nothing new under the sun, so the observation that authors follow a consistent pattern is scant criticism, but it is worth noting the pattern points towards a common response to the Lynn White hypothesis from the liberal theological perspective for environmental ethics.

In general, liberal theologians accept White's accusation that Christianity is responsible for the present environmental problems. Those problems, they believe, are so significant that even dearly held traditional beliefs should be subject to questioning, at least, to determine if the beliefs form an adequate foundation for a Christianity that confronts and counters the ecological crisis.[40] The pattern the aforementioned authors follow, therefore, is designed to support the need for potential revision for the common good.

There are at least two reasons that the White hypothesis has generally been accepted by liberal Christians. The first is that White's call to modify Christianity in light of a contemporary criticism resonates with the accomodationist attitude that seems to permeate liberal Christianity.[41] Thus, liberalism has room to modify Christianity based on White's criticism just as it has modified traditional forms of Christianity based on criticism from the scientific community in the past, all while trying to maintain the essential kernel of Christianity.[42]

A second reason that liberal Christians have generally accepted White's criticism is that it is largely correct for modernistic forms of Christianity. When White claims that the desacralization of nature results in the devastation of the environment through human abuse, this is largely due to the rise of modernistic science, which resonates in many ways with the humanistic origins of liberal Christianity.[43] In *Critique of*

---

39. Kureethadam, *Creation in Crisis*, 329–73; McDonagh, *To Care for the Earth*, 143–214; Nash, *Loving Nature*, 139–221; Northcott, *The Environment and Christian Ethics*, 165–327; Rasmussen, *Earth Community Earth Ethics*, 322–54.

40. Fowler, *The Greening of Protestant Thought*, 61–64.

41. Carson notes that theological liberals clearly fall within the "Christ of Culture" among Neibuhr's category, thus adopting an accomodationist model. Carson, *Christ and Culture Revisited*, 33.

42. Machen, *Christianity and Liberalism*, 6.

43. See, for example, the understandings of science as the pursuit of absolute mastery of nature, which Gillespie finds in both Descartes and Hobbes. It is this modernistic attitude that has led to the devastation of nature, not traditional Christian doctrine. However, as discussed above, there are ties between modernity and liberal Christianity, which help explain the sense of guilt among liberal Christians over the

Modernity, Alain Touraine notes that modernism has itself been forced to completely flip its perspective on human control of the environment:

> The concern for the environment and the increasing importance of ecologist parties is an even more spectacular demonstration of how ideas and feeling have been inverted ... Early modernity ... celebrated man's dominance over nature. The tendency is not to agree with the scientifically trained ecologists who insist that, whilst human beings do have the ability to transform the world, they must also take in to account the various effects their actions have on even the most distant parts of the system.[44]

Since liberal Christianity has its roots in modernism, and modernism initially exalted human dominance over creation, it stands to reason that liberal Christians would need to repudiate their own traditions in order to stand within the contemporary environmental movement. Thus, White's thesis is partially correct; there is a close tie between the rise of modernistic liberal Christianity and the environmental crisis.[45] This correspondence helps explain why many in the liberal theological perspective for environmental ethics embrace White's thesis and accept the need for revisions to Christian theology.

According to Fowler, some critics within Christianity find their religion at fault "because the Renaissance and the Reformation created no profound reformulation of Christian doctrine regarding nature. Moreover, the secular offshoots of these great periods, such as modern science and technology, have too often proved pernicious [to the environment]."[46] Since earlier forms of Christianity are held to be implacably dualist by some theologians, the failure to reformulate the human-creation relationship is detrimental in the eyes of modern environmentalists. One of the troubles with this sort of blame-throwing is that there is a great deal more data on the ideas of theologians of Christian history and much less on ecological attitudes and lifestyles of many Christians. As Fowler notes, "Taoism and Buddhism are profoundly sympathetic to the ideal of ecological wholeness. Yet China's natural environment has suffered greatly over the centuries."[47] Ideas often have consequences, but prevailing ideas

---

environmental destruction. Gillespie, *The Theological Origins of Modernity*, 222–34.

44. Touraine, *Critique of Modernity*, 301.
45. Spencer, "The Modernistic Roots of Our Ecological Crisis," 355–71.
46. Fowler, *The Greening of Protestant Thought*, 65.
47. Fowler, *The Greening of Protestant Thought*, 71.

may be significantly different from the ones recorded for later analysis. In the end, it may be proved that significant revision of Christian doctrine along ecological lines has been unwarranted. The roots of the present ecological crisis seem to be more complicated than White allows.

White is a medieval historian who specialized in the history of technology. However, other medievalists have come to significantly different conclusions regarding the role of Christianity, properly speaking, in the desacralization of creation. The Christian pursuit of science was not a move away from sacredness in the created order, but it was an attempt, just as with magic, to subdue order and bring nature to heel. C. S. Lewis—who well knew the literature of the Middle Ages—writes,

> There is something which unites magic and applied science while separating both from the "wisdom" of earlier ages. For the wise men of old the cardinal problem had been how to conform the soul to reality, and the solution had been knowledge, self-discipline, and virtue. For magic and applied science alike the problem is how to subdue reality to the wishes of men: the solution is a technique; and both, in the practice of this technique, are ready to do things hitherto regarded as disgusting and impious—such as digging up and mutilating the dead.[48]

Both magic and science result in the degradation of nature in an attempt to control it. Lewis's "wise men of old" likely refers to thinkers in the Western tradition, including many Christians. His account places the fault for desacralization outside of the Christian tradition.

White's recommendation for the development of an eco-friendly Christianity is to adopt Saint Francis of Assisi as a model for a Christian attitude toward the environment and also to accept a pantheistic conception of nature.[49] Francis of Assisi has become a popular figure for some liberal Christian environmentalists.[50] However, the movement toward pantheism or panentheism has not taken hold within most streams of Christian liberalism.[51] Instead of moving all the way to pantheism, liberal Christians have developed alternative emphases within their theologies to enhance eco-friendliness.[52]

48. Lewis, *The Abolition of Man*, 77.
49. White, "The Historical Roots of Our Ecological Crisis," 134–37.
50. For example, see Santmire, *The Travail of Nature*, 107–19. See also Fowler, *The Greening of Protestant Thought*, 59.
51. One exception is Van Wieren, *Restored to Earth*, 78–79.
52. Fowler, *The Greening of Protestant Thought*, 69.

Revelation

A step toward establishing those alternative emphases in liberal theology has been adopting an approach to Scripture suitable for doctrinal modifications. Revision of traditional Christian positions must begin by establishing flexible boundaries for the interpretation of Scripture. The views of the nature of revelation held by theologians in the liberal theological category range along a spectrum. At one end there are those who hold a high view of Scripture and its authority for life as it is applied through thematic interpretation.[53] At the other end of the spectrum are those theologians who hold a lower view of Scripture, seeing it as reflecting the cultural prejudices of the human authors.[54]

The doctrine of revelation is one of the defining attributes of liberal theology. Robert Fowler uses attitudes toward special revelation, especially Scripture, as the defining attribute of liberals for his survey of Protestant environmentalism: "I define more liberal Protestants as those who employ the Bible as an important part of their faith, as a document that is inspired by God but that is also a historical and cultural work. Liberal Protestants do not contend that the Bible is somehow the full truth of God or that it is true in all aspects."[55]

Alvin Plantinga labels a liberal approach to reading Scripture "historical biblical criticism" (HBC) as opposed to "traditional biblical scholarship."[56] According to Plantinga, HBC is "fundamentally an Enlightenment project. It is an effort to look at and understand biblical books from a standpoint that relies on reason alone."[57] It is an attempt to abandon presuppositions derived from "the authority and guidance of tradition, creed, or any kind of ecclesial or 'external' epistemic authority."[58] Plantinga argues that "practitioners of HBC like to wrap themselves in the mantle of modern science."[59]

---

53. Gustafson, "The Place of Scripture in Christian Ethics," 163–64.

54. For example, McDonagh argues specifically for the flawed nature of the masculine characteristics of God in Scripture: McDonagh, *To Care for the Earth*, 114–17.

55. Fowler, *The Greening of Protestant Thought*, 5. He gives more detail in Fowler, *The Greening of Protestant Thought*, 32–33.

56. Plantinga, *Knowledge and Christian Belief*, 97.

57. Plantinga, *Knowledge and Christian Belief*, 97.

58. Plantinga, *Knowledge and Christian Belief*, 97.

59. Plantinga, *Knowledge and Christian Belief*, 97.

Within the category there are two sub-categories, which reach similar conclusions about the content and nature of Scripture, but apply somewhat different methods to historical biblical criticism. Plantinga labels the first Troeltschian HBC, which is named after the nineteenth-century German philosopher, Ernst Troeltsch. To adopt a Troelschian HBC, "you have to assume that God has never acted directly in the world."[60] To the theologically trained ear this sounds more like deism than Christianity, but it reflects the idea of creation as a closed system, which is implicit in a liberal approach to Scripture. Plantinga calls the second category of historical biblical criticism Duhemian HBC, which is named after Pierre Duhem, a Roman Catholic scientist. Duhem tended toward agnosticism about the supernatural. He neither assumed that God did act in history, as for example by raising Christ from the dead, nor did he assume that God necessarily did not act in history. The goal of a Duhemian approach is to evaluate the contents of reality, including the witness of Scripture, based on evidence upon which everyone can agree so that "people of very different religious and theological beliefs can cooperate" in the effort.[61]

Plantinga's taxonomy is helpful, because any careful study of biblical interpretation within the liberal tradition will uncover a range of attitudes, of which Plantinga's sub-categories represent closely related but distinct approaches. In both cases, the biblical critics still generally hold the text to be authoritative in what it discusses, but they would tend to relegate much of what Scripture says to the human authors' cultures, with potential or actual errors embedded in the text. The issue for theological liberals is not that Scripture is no longer important, but that the accepted findings of science and influence of culture are weighed more heavily than in conservative theological perspectives. Unlike fundamentalists and conservative evangelicals, theological liberals are willing to accept the existence of errors in Scripture.[62] The central common characteristic is the desire to approach the text "scientifically" as a means of giving contemporary social consensus a place when reading Scripture.

Another way to express the understanding of the nature of Scripture as a form of revelation for liberal theologians is to evaluate the relative importance of science in the Wesleyan Quadrilateral. "Real liberal

---

60. Plantinga, *Knowledge and Christian Belief*, 99.

61. Plantinga, *Knowledge and Christian Belief*, 100–101.

62. According to Olson, a central aspect of classical liberal Protestant theology is that "the Bible is recognized as the primary Christian classic but not as supernaturally inspired or infallible." Olson, *The Journey of Modern Theology*, 129.

theology aims at reconstructing Christian doctrines to balance contemporary cultural relevance with faithfulness to Christian sources," argues Olson. "Usually, and this is probably the *sine qua non* of liberal theology, relevance to contemporary culture [including contemporary scientific data] is given equal if not greater weight than faithfulness to Christian sources."[63] This approach, which is essential to theological liberalism, enables conditions in which scientific consensus can drive ethical responses within liberal Christianity that are contrary to the precepts of Scripture.

There are a number of attitudes toward contemporary science among theological liberals. Olson details three: (1) acquiescence to science over Scripture; (2) accommodating science by granting it a parallel authority over life in areas not explicitly governed by Scripture; and (3) attempting to integrate scientific facts into Scripture, though in this approach scientific data appears to have the pride of place.[64] For the truly modern thinker, experimental science, which is the embodiment of reason, is the primary source for theology within the quadrilateral. Fowler notes that in some cases, particularly in theologically liberal environmentalism, "Science is often taken as the central authority. The irony is that in a world where idolatry of science is often attacked, there are some signs that it is also practiced."[65] A comprehensive evaluation of the complex relationships between science and Scripture remains beyond the scope of this project, but at least in some cases, for liberal theologians, data from science has a role in ethical decision making that supersedes the authority of Scripture.[66] This acceptance of scientific authority often relies on a diminution of the noetic effects of the fall on the human mind.

As described by Langford, the impact of the fall on human reason is limited. Langford argues the liberal theologian will tend to value the influence of contemporary reason and experience more highly than conservative theologians.[67] Langford specifically opens up the door for authentic, authoritative revelation of God coming through external sources such as the scriptures of other religions. He claims that a high view of Scripture allows critiquing the Bible in the same manner as other literature. He allows the "unique authority" of the Bible while ignoring

63. Olson, *The Journey of Modern Theology*, 128.
64. Olson, *The Journey of Modern Theology*, 43–44.
65. Fowler, *The Greening of Protestant Thought*, 5.
66. For a more thorough treatment of the relationship between science and religion, see McGrath, *The Foundations of Dialogue in Science and Religion*, 1–29.
67. Langford, *The Tradition of Liberal Theology*, 24–32.

the possibility it has a unique character.[68] In other words, for Langford, who is claiming to represent other liberal theologians, the Bible has pride of place but not categorical distinction. This approach results in the blurring of general and special revelation.

When the distinction between general and special revelation is diminished, influences such as prevailing interests of science, politics, and art will tend to have a more significant role in driving theological interests and interpretations of Scripture. All information begins to morph into a constant stream of knowledge of God, however defined. Thus, even a liberal theologian who holds a high view of Scripture will tend to find more references or allusions to the culture's main topic of interest. This helps to explain why some environmental ethicists who demonstrate a relatively high view of Scripture will begin to read more ecological implications from Scripture than other, more conservative theologians will.[69]

For example, Michael Northcott and Peter Scott, in their introduction to *Systematic Theology and Climate Change*, make a bold assertion along these lines: "That new contexts call forth new teachings is for Christian theologians underwritten by the Trinitarian belief that the Holy Spirit, who was gifted to the church at Pentecost, continues to reveal new truth as Christians reflect on scripture and tradition in new contexts."[70] The first portion of this statement is not controversial even to the most traditional of theologians—new contexts require new teachings—but the assertion that the Holy Spirit continues to reveal new truth reflects a different understanding than the attitude most evangelicals and fundamentalists display toward the canon of Scripture.[71] A theory of ongoing inspiration of revelation from God with authority equal to or perhaps greater than that of Scripture enables the very revisionism of doctrine that defines liberal Christianity. It also allows a heavier reliance on sources other than canonical Scripture for theological ethics.

James Gustafson presents his understanding of the sources of theology for ethics in clear, picturesque language, stating,

---

68. Langford, *The Tradition of Liberal Theology*, 23.

69. Fowler, *The Greening of Protestant Thought*, 31–37.

70. Northcott and Scott, "Introduction," 1.

71. This discussion will neglect the charismatic beliefs of some Christians who label themselves as either "evangelical" or "fundamentalist." In their best forms, even those believers who allow ongoing special revelation tend to assert the preeminence of canonical Scripture. For example, see Williams, *Renewal Theology*, 22–25.

> Perhaps the best metaphoric description of the technical structure of the position [on the sources of authority for theology] is that of a raft, as suggested by John P. Reeder. Like the Coca-Cola sign fishing raft we used on Brown's Lake in Dickinson County, Michigan, there are four barrels, one in each corner: evidence from relevant sciences (and never "science" as one reified whole), human experience, philosophical judgments, and the heritage of Christianity and Judaism as continued in those traditions. All are necessary, and none is sufficient, but the raft tips to one corner or another depending upon the weight of the exposition and defense.[72]

By maintaining the necessity of all four sources, which map quite well onto the so-called Wesleyan Quadrilateral, Gustafson is able to claim his theocentric perspective on ethics is "deeply informed by the Bible and traditions which flow from it."[73] Meanwhile he still retains flexibility within his ethical schema to adapt Christian dogma to contemporary scientific concerns as deemed necessary. It also leaves a wide entryway for bringing personal experience and emotion in as sources for environmental ethics.

While science, or at least some forms of science, is clearly influential in contemporary liberal approaches to environmental ethics, emotive response bears a part in that as well. This is consistent with the experiential-expressivist understanding of revelation, which is characterized by Friedrich Schleiermacher, arguably the father of theological liberalism.[74] For Schleiermacher, the Bible is not in itself revelation from God but a response that faith awakened within Scripture's authors. Thus, in reference to the authors of the Gospels, Schleiermacher argues, "Their description of Jesus was only an expression of this faith combined with their faith in the prophets."[75] Thus Schleiermacher excludes belief in the facticity of Scripture from the list of necessary doctrines of a Christian because the Bible is a contextually bound, imperfect communication of personal experience. The doctrine of Scripture is of interest as a part of revelation mainly because it has been tied to the Christian tradition and reflects the early record of responses to faith.[76] In his introduction, Schleiermacher essentially eliminates from consideration the possibility of the authentic

---

72. Gustafson, *A Sense of the Divine*, 46.
73. Gustafson, *A Sense of the Divine*, 46.
74. Olson, *The Journey of Modern Theology*, 73–74.
75. Schleiermacher, *The Christian Faith*, 592.
76. Schleiermacher, *The Christian Faith*, 594–96.

communication of revelation from one human to another, as through Scripture.[77] Schleiermacher is left with a need for the individual to receive revelation from God directly and that through a sense of absolute dependence on God. External authority is diminished in reality, though this position maintains a strong sense of authority for the individual's personal experience. Schleiermacher thus clears the way for accepting contemporary scientific understandings and human experience as sources of authority like or exceeding that of the canon of Scripture.

Even within liberal Christian perspectives that are more rational and less emotive than Schleiermacher's, there is space for emotion to help steer ethical reasoning. Gustafson's "theocentric perspective informs and empowers a sense of radical dependence and a consciousness of the ambiguities of the many relations of multiple values of things for each other in nature."[78] The sense of the divine in nature can be found through experience because, Gustafson insists, "We meet God through nature, as well as through historical and interpersonal human events. We meet God as the power that brings all things into being, that bears down on them and threatens and limits them, that sustains them and is the condition of possibility for their change."[79] The experience of the divine through exposure to nature creates a response within the individual that helps create the beginning of environmental ethics.

As Robert Fowler notes, Christian environmental ethics texts typically begin with a personal account of experience with nature and scientific data that illustrate the problem.[80] The emotional response often approximates what evangelicals would describe as a conversion experience, but it is a conversion to environmentalism. This emotional response is the foundation for the vision of the arguments that come later; respect or awe of nature provides the presuppositional basis that drives exegesis of the Scripture. Scientific data is accepted, in part, because it resonates with the emotional response (or perhaps caused it) and also provides a basic motivation for environmental concern.

77. Schleiermacher, *The Christian Faith*, 51–52.
78. Gustafson, *A Sense of the Divine*, 48.
79. Gustafson, *A Sense of the Divine*, 14.
80. Fowler, *The Greening of Protestant Thought*, 21.

Creation

The emotional response to nature itself, or the sense of divine derived through exposure to creation, is undergirded by a particular understanding of the doctrine of creation. In particular, the emotional response to the sense of the divine in nature relies on the immanence of God in nature and the acceptance of an ongoing relationship between the Creator and the creation. In general, for liberal environmentalists, a key factor in trying to redefine Christianity as an ecofriendly religion has been the resacralization of nature.

As Norman Wirzba argues in his recent book *From Nature to Creation*, the move to understand the sacredness of the created order is a reaction to modernism, which tended to reduce the world to "mundane matter in motion."[81] He states, "Though ancient Greek and Roman thinkers were often anthropocentric in their understanding of humanity's place in the world, placing humans highest among all creatures, modern thinkers took the new step of lifting human beings out of creation altogether so as to rule over it in ways they saw fit."[82] As Wirzba writes, this attitude was carried out by individuals who

> continued to invoke the name of God, but the god they referred to bears little resemblance to the Creator as proclaimed in Psalm 104, who sends forth the creative spirit/breath that daily renews the face of the ground, or the Triune God who is intimately and constantly present to the world as its sustaining, beautifying, and perfecting end.[83]

Wirzba seems intent on inspiring his readers to a greater sense of imaginative wonder at God's good creation by attempting to reinvigorate the sense of God's connectedness with the created order.

However, establishing wonder and a sense of divine purpose in creation is complicated within the liberal Christian theological perspective for environmental ethics by the dominance of modernistic scientific data as a theological source. The same naturalistic, scientific outlook that enables rejecting the veracity of miracle accounts within Scripture also helps to undermine a doctrine of creation that supports environmentalism. In other words, it is difficult to envision a God so united with his

---

81. Wirzba, *From Nature to Creation*, 14.
82. Wirzba, *From Nature to Creation*, 16.
83. Wirzba, *From Nature to Creation*, 16.

creation that it has intrinsic value, yet so distant that he is incapable or unwilling to dynamically interfere through miracles or, even more basically, through the process of creation by fiat instead of naturalistic evolution. On the other hand, there is a sense of community of human and non-human creation that can be drawn from accepting an evolutionary theory of origins, which impacts the value structure.

For liberal Christians, the idea that the earth was indeed created in a seven-day period only a few thousand years ago is so far out of line with the prevailing scientific theories that it is not considered a viable interpretation of Genesis 1.[84] Therefore, the doctrine of creation does not begin with cosmogenic theory but with the nature of the ongoing creative work of God.[85] The assumption is that the universe is billions of years old and that human existence is a mere speck on the grand tapestry of universal history. This serves, for Joshstram Kureethadam, as the foundation of his environmental ethics: creation has existed so much longer than human history that it must, logically, have value consistent with or superior to humanity's.[86] Rather than denigrate human value, Kureethadam and others tend to elevate the value of the non-human creation.

Toward that end, Kureethadam argues that establishing the connection of humanity with the earth is necessary for a proper environmental ethics. He states that one of the most significant problems with Christian environmentalism is that "we have not discovered ourselves as inhabitants of the common home of earth, as truly earthlings, *imago mundi*, formed from the dust of the earth."[87] At the same time he also makes an argument regarding God's connection with the created order. Thus he sees the ecological crisis as being caused, in part, by "our failure to recognize God's in-dwelling presence in creation, a presence that renders it God's own home. God is not only transcendent to creation but is also deeply immanent in it."[88] He grounds this connection in both a Trinitarian conception of the creation event and the incarnation of Christ. This grounding keeps Kureethadam within the boundaries of Roman Catholic orthodoxy, but the language he uses illustrates Olson's description of pantheistic themes within liberal Christian theology. Kureethadam and

84. E.g., Cox, *How to Read the Bible*, 26; Nash, *Loving Nature*, 96–97.

85. E.g., McDonagh, *To Care for the Earth*, 117–19; Wirzba, *From Nature to Creation*, 20–21.

86. Kureethadam, *Creation in Crisis*, 65.

87. Kureethadam, *Creation in Crisis*, 49.

88. Kureethadam, *Creation in Crisis*, 300.

many others who build arguments about the value of the created order on God's immanence in the created order are not proposing a pagan-like worship of the earth. Instead, they are building the case for the value of creation by identifying creation more closely with God. This is often enhanced by emphasizing Christ's connection with the physical creation in his incarnation.

Christological approaches to environmental ethics, far from presenting a challenge to the four doctrinal emphases in a theological perspective on the environment, tend to reinforce the validity of the importance of the doctrine of creation. In his explication of a Christology for climate change, Niels Henrik Gregersen finds support for environmentalism because of the close relationship between God and the creation in the incarnation of Jesus Christ. He argues, "Christology is therefore not first and foremost a backwards-oriented remembrance of the teachings of Jesus in his earthly life nearly two thousand years ago. Christology is carried by the conviction that God's eternal Logos has revealed and re-identified itself—once and for all—*as* Jesus Christ *within* the matrix of materiality that we share with other living beings."[89] Thus within a liberal ecological theology, Christology often becomes a means to emphasize the immanence of God in creation rather than a distinct theological theme of its own merit. By drawing God into solidarity with creation consistent with orthodox theological themes, the value of creation can be elevated without resorting to pantheism or panentheism, though the typical vocabulary used sometimes becomes problematic.[90]

Many liberal environmentalists ascribe intrinsic value to the created order or a part of the created order.[91] For example, Kevin O'Brien argues that the "core idea of intrinsic value is at the heart of many arguments on behalf of conservation."[92] "Intrinsic value" strictly speaking refers to something having value in and of itself. However, O'Brien modifies the meaning of "intrinsic value," instead suggesting "that if something is intrinsically valuable, its value does not depend on human interests; we do not judge its worth based on its worth to ourselves."[93] By changing the

89. Gregersen, "Christology," 37. Emphasis original.

90. A notable exception to this is Lutheran theologian Larry Rasmussen, who claims to find panentheism in Luther's writings and builds his environmentalism on this. Rasmussen, *Earth Community Earth Ethics*, 273–79.

91. E.g., Nash, *Loving Nature*, 132.

92. O'Brien, *An Ethics of Biodiversity*, 53.

93. O'Brien, *An Ethics of Biodiversity*, 51.

definition, O'Brien manages to avoid the danger of pantheism, but he uses the term imprecisely. His language resonates with the deep ecology movement, but his meaning is not consistent with theirs.

On the other hand, James Gustafson presents a balanced understanding of the value of creation. He recognizes the physical ability of humans to intervene in natural processes and also allows situations in which humans should intervene.[94] He calls for humans to value aspects of nature due to something other than its instrumental value. Gustafson argues for a position between two poles, insisting,

> One extreme is that nature has intrinsic value or rights, that nature itself is sacred, that our attitude toward it ought to be reverence. The other extreme was utilitarian, that nature is there for our use and benefits as human beings . . . [F]rom a theocentric perspective nature evokes in us a sense of the sublime, or more religiously, a sense of the divine, [which] calls for *respect* for nature.[95]

Gustafson communicates the position that most liberal environmentalists intend with his language. It is not that nature has value in and of itself apart from God, rather it is that nature is imbued with value by God because he created it.

Gustafson's understanding of value is consistent with many theological liberals, though the language of nature having intrinsic value, as O'Brien argues, is more prevalent.[96] The resonance between the language of value for the deep ecology movement and many liberal Christian environmentalists enhances opportunities for co-belligerence, but it also confuses categories of value and opens theological liberals to accusations of pantheism or panentheism.[97] Improving the clarity of the terms used would go a long way toward enhancing dialogue, but the point remains that in general, what many liberal theologians mean by arguing for intrinsic value of nature can be reconciled with orthodox Christian theology.

---

94. Gustafson, *A Sense of the Divine*, 103.

95. Gustafson, *A Sense of the Divine*, 55. Emphasis original.

96. O'Brien, *An Ethics of Biodiversity*, 51–53. For example, Nash uses the word *instrinsic* to describe the value of nature, but his explanation of the concept in the doctrine of creation sounds more like *inherent* value. Nash, *Loving Nature*, 95–100.

97. For more on how this has confused the discussion, see the helpful discussion on alternate value theories in Sarkar, *Biodiversity and Environmental Philosophy*, 45–105.

It does, however, have an impact on the doctrine of anthropology as it relates to human responsibility within the created order.

Anthropology

One of the key tenets of theological anthropology within the liberal Christian environmental paradigm is that humanity is merely part of nature.[98] The hierarchical distinctions between human and non-human creation are erased, or at least significantly blurred. Anthropocentrism is the bogeyman that must continually be combatted within the human heart, even as it is uncovered in Scripture.[99] Evolutionary theory has enabled a flattening of the classical understanding of the created order. Instead of seeing humanity as a special part of creation, even the pinnacle of God's work in creation, there is a tendency within liberal Christian environmental ethics to see humans as merely a product of the natural processes of evolution. When viewed through this lens, it appears inconsistent with the value of creation for humans to interfere with ecological processes and change natural patterns, whether with intent to improve the human condition or merely due to careless abuse.

At the same time, given liberalism's roots in modernity, there is little obvious reason for humans to work for the benefit of nature at their own expense. Both Descartes and Hobbes were instrumental in reforming science and directing it toward ends beneficent to humanity. However, their anthropology is significantly different. Both deny the existence of universals, which means that science was redirected from the purpose of understanding God better through knowledge of the world to that of subduing nature to reduce human suffering. Descartes recognizes both a material and immaterial aspect to humanity, while Hobbes "rejects the idea that humans have a supernatural component as a ploy of priests to gain power over others."[100] Gillespie argues that like Descartes, "Hobbes too seeks to make man master and possessor of nature, but in contrast to Descartes, he denies that human beings have any special status."[101] These two interlocutors are immensely significant in the development of modernism and therefore liberal Christianity. The work of liberals to redeem

---

98. Scott, "Humanity," 109.
99. E.g., Rasmussen, *Earth Community Earth Ethics*, 187–92.
100. Gillespie, *The Theological Origins of Modernity*, 234.
101. Gillespie, *The Theological Origins of Modernity*, 222.

Christianity from its anthropocentric roots is an effort to distance itself from the formulations of Hobbes and Descartes that had influence on the modernism from which liberalism arose, including the dualism that Descartes espoused.

In contemporary environmental writing, there are few insults more severe than accusing an individual of dualism, particularly Cartesian or substance dualism. In fact, Bradley Green, in his dissertation-turned-monograph, spends just over two hundred pages defending Augustine from the accusation of being a hard dualist.[102] Some degree of dualism appears to be a basic element of Christian orthodoxy.[103]

There seems little reason for a Christian environmentalist to reject the notion that there is a material and immaterial aspect to humanity, and that the normal condition of humans is for these two substances to remain together, both in this life and in the resurrection at the end of the age. But some forms of dualism come with dangers. As Luke Stamps has recently shown, the difficulty with Cartesian dualism compared with other versions is the assertion that the human person *is* the soul.[104] From that understanding a Cartesian dualism could promote a neglect of the material world by minimizing the significance of the human body and thereby the physical world, even while not promoting the anti-material perspective of gnostic dualism.

One reason for accepting a sort of monism or a more carefully delineated biblical dualism is that it allows for accepting humanity as an undifferentiated part of creation; it weakens the idea that the *imago Dei* is best evidenced by human mastery over nature and enables the acceptance of humans as merely evolved animals. Even accepting the distinction of humanity as the highest and most rational form of living creature, this still places humans directly within the ecological web and tends to give humanity the responsibility to serve the non-human creation instead of ruling or subduing it.

Indeed, one of the most significant passages of Scripture for a Christian environmental ethics is Genesis 2:15. Noted revisionist environmentalist H. Paul Santmire re-envisions the meaning of Genesis 2 significantly.[105] He writes,

102. Green, *Colin Gunton and the Failure of Augustine*.
103. Deweese and Moreland, *Philosophy Made Slightly Less Difficult*, 105–07.
104. Stamps, "A Chalcedonian Argument Against Cartesian Dualism," 53–66.
105. Santmire describes his project as a "revisionist theology, which is methodologically bound to take the witness of the Scriptures with utmost seriousness."

This earthly creature is then commanded to serve and to protect its earthly home. This is no command to exploit, although it has often been interpreted this way. Rather it is a command to care for the garden, presumably by being attuned to the needs of the garden, as any good gardener was surely understood to be in biblical times as in our own. Phyllis Trible has pointed out that the Hebrew word *'bd* is appropriately translated "to serve," rather than the more conventional "to till": "it connotes respect, indeed, reverence and worship." Likewise the word *smr*, usually translated "to keep," is much more adequately rendered "to protect."[106]

There appears to be little debate prior to the mid-twentieth century on the appropriate translation of Genesis 2:15. However, since that time, which corresponds roughly with the rise of the movement toward uniting Christian theology with a concern for ecology, this translation has become increasingly popular.

Even the potentially eco-centric ethics that could result from such a retranslation of Genesis 2:15 are very different from neo-paganism or worship of the earth. This understanding of serving creation still recognizes the unique capacity for human interaction with nature and allows for well-reasoned human interventions in nature. Within texts on environmental theological ethics written from a liberal perspective, it is more common to see the obligation to serve creation assumed rather than defended from Scripture. Or, the question of the translation can be dealt with by placing two apparently contrasting passages of the Bible together and privileging the authority of one passage over the other. For example, Sean McDonagh skirts the issue, saying, "It is true that in Genesis 1 and 2 the human is given a special position, but some scripture scholars, like Bernhard Anderson, insist that to focus exclusively on the first two chapters of Genesis and leave chapter 9 out of the reckoning presents a very unbalanced picture of the biblical notion of creation."[107] Whether and however Genesis 2:15 is dealt with within the liberal perspective, a common result is an understanding of the human role in creation that is not entirely anthropocentric or eco-centric.[108]

Santmire, *Nature Reborn*, 31.

106. Santmire, *Nature Reborn*, 39–40.

107. McDonagh, *To Care for the Earth*, 122.

108. Calvin Beisner argues that translating Genesis 2:15 as "serve and protect" opens the door for an eco-centric theology, but recognizes that such an outcome is not

Gustafson well represents a balanced approach to human engagement with creation that recognizes both the distinction of humanity from non-human creation and the connection of humans with the rest of the created order, writing,

> Human beings participate in the patterns and processes of interdependence of life in the world. We can and should intervene for the sake of humans and nature itself. Our participation is a response to the events and conditions in which we live; it involves valuing aspects of nature in relation not only to our own interests but also the "interests" of other aspects of nature. Human life is dependent not only upon the processes and patterns of nature, but also upon human intentional participation in them for the sake of survival and other justifiable values to the human species, communities, and individuals. It takes seriously our biological histories; we have come to be as a result of processes of nature in the cosmos, in the development of this planet, and the development of mammalian species, etc.[109]

Gustafson relies on an appreciation of the value of nature apart from instrumental value for humans, but does not cross the line into divine worth of creation. His position also recognizes a distinct human rule in creation, so that humanity sometimes intervenes for nature itself. In other words, for Gustafson, there is still a place for humans as stewards of God's creation; all human impacts on the created order are not a result of sin.

Though Kureethadam takes great pains to minimize the significance of humanity within nature with his extensive cosmogony, he too has a high view of the human role in creation. He argues, "In being 'co-carers' of God's creation, humans are to imitate and reflect God's own tender and loving way of caring for the physical world."[110] He finds this responsibility rooted within the very nature of humanity as the *imago Dei*.[111] Sinful human interaction with the environment becomes "any human action that damages our common home and endangers the life and survival of our common household."[112] This perspective respects the unique role of humans within the created order, but it does not provide a clear boundary

---

necessary. Beisner, *Where Garden Meets Wilderness*, 15–18.

109. Gustafson, *A Sense of the Divine*, 103.
110. Kureethadam, *Creation in Crisis*, 332.
111. Kureethadam, *Creation in Crisis*, 332–33.
112. Kureethadam, *Creation in Crisis*, 338.

to determine what damage is and at what point human impact for human benefit crosses the line into sin. This ambiguity opens the liberal theological perspective to criticisms of misanthropy and resistance to economic development.

Eschatology

Though it is never eliminated, the ambiguity of the exact nature of human involvement in the created order is clarified somewhat by examining the eschatological vision that accompanies a liberal theological perspective for environmental ethics. Unlike the fundamentalist and evangelical theological tradition, there is little emphasis among liberals on the question of the millennium, unless it is to criticize fundamentalists and evangelicals for being concerned with the nature of the millennium.[113] The move to moralize dogma, as Olson describes it, is particularly clear in the treatment of eschatology from the liberal perspective on environmental ethics.

Stefan Skrimshire offers this outline of his understanding of eschatology for environmental ethics, writing: "Eschatology concerns that in which a believer may legitimately place their hope, from purely personal destiny (life after death) to the fate of the cosmos and the 'end of all things'. It is also the doctrine that is most likely to have direct impact on the direction of political action, generating attitudes that range from resignation and apathy to triumphalism in the face of crisis."[114] Eschatology is inextricably tied to the concept of redemption. According to Larry Rasmussen, "The object of redemption is to free people and the rest of nature to become what all was created to be."[115] For Rasmussen there is little need for a personal eschatology because death is necessary and should not be avoided. Instead, "Our problem is not that we are mortal but that we are unjust."[116] Rasmussen pushes the boundaries of the liberal category to the left by his promotion of panentheism, but his focus

---

113. Northcott, "The Dominion Lie," 94–100.
114. Skrimshire, "Eschatology," 158.
115. Rasmussen, *Earth Community Earth Ethics*, 256.
116. Rasmussen, *Earth Community Earth Ethics*, 277. Nash helps to explain this by noting that death is not an issue for those that have accepted that life is merely a biological accident. Nash, *Loving Nature*, 130.

on cosmic redemption as the primary eschatological theme is consistent with the liberal theological perspective on the environment.

Doctrine should be eminently practical from the liberal perspective. Thus, for Nash, the resurrection need not be an actual event, but it is a necessary theme that points toward the full redemption of all creation.[117] Hope can be found in the theme of the resurrection—in the idea that God identified with creation closely and conquered death to bring about a future restoration of all things. But the particulars of the doctrine of atonement, including its necessity due to human sin, disappear entirely from Nash's account of the crucifixion and resurrection.[118] This opens up new questions regarding salvation for individuals, the importance of pursuing holiness on earth, and the potential of a divine judgment. It also hearkens back to the theological concept of universal salvation, which Olson claims is central to liberal theology. If redemption is for all creation without differentiation, as Nash seems to propose, then universal salvation is not only possible but absolutely necessary.

Michael Northcott associates the modern belief in the progress of human society with a particular eschatological vision that is futuristic in nature.[119] For Northcott, "the recovery of an ecological ethic in the modern world requires the recovery of a doctrine of creation redeemed, and the worship of a creator who is also the redeemer of the creation."[120] For James Nash the concept of God's redemption of creation gives meaning and value to the created order.[121] The eschatological vision of a redeemed creation "represents the ultimate goal to which God is beckoning us. Our moral responsibility, then, is to approximate the harmony of the New Creation to the fullest extent possible under the constricted conditions of the creation."[122] In other words, from the liberal theological perspective, eschatology defines the *telos* of the created order and thus points toward the way that Christians should live in the present.

In a discussion of sin and salvation, which are inextricably linked to eschatology, Neil Messer argues,

---

117. Nash, *Loving Nature*, 131.
118. Nash, *Loving Nature*, 129–33.
119. Northcott, *The Environment and Christian Ethics*, 67.
120. Northcott, *The Environment and Christian Ethics*, 222.
121. Nash, *Loving Nature*, 132–33.
122. Nash, *Loving Nature*, 133.

> The crisis of climate change is the sharpest of reminders that humanity's fate is bound up with that of the whole created earth that we inhabit. As such it can redirect the attention to aspects of salvation and sin in the Scriptures that Christian reflection has too often neglected; it can prompt us to give an account which acknowledges our solidarity with the whole of creation, and remind us of the biblical witness to a divine purpose of transformation and renewal that is not limited to our species.[123]

Thus personal eschatology and hope are motivations to be an environmental activist and establish so-called green practices in a congregation. Personal holiness is evidenced through green living. Hope is found in the redemption of all things rather than the atonement of one's particular sins.[124]

While Nash and Northcott seem to hold out hope for a literal cosmic redemption, Skrimshire's eschatology is less futuristic. He writes, "The world is not eternal: this belief is fundamental to orthodox Christian thought. But if theology is to articulate this in a way that is meaningful to the activist, it may need to be through the lamenting—and resisting—of future *failures* within the time allotted to us, not the utopian, consoling vision of some future one."[125] This brings eschatology down to the level of a self-deception to enliven immediate political effort to minimize damage to the environment.

The liberal perspective on eschatology is diverse, much like on the other three doctrines. However, the consistent theme in liberal eschatology is that the purpose of the doctrine is to inspire ethical action that, within the present context, includes ecological activism. Whether Scripture refers to a literal or merely figurative restoration of all things, redemption of all creation is the end toward which Christian ethics should point and toward which Christians should hope. Anything that falls short of that or goes beyond that is folly. As Nash argues, "Despite the hordes of speculations and imaginative descriptions of glory in Christian history, an honest and humble Christianity knows when to keep silent."[126]

---

123. Messer, "Sin and Salvation," 130.
124. Messer, "Sin and Salvation," 134–36.
125. Skrimshire, "Eschatology," 168.
126. Nash, *Loving Nature*, 132.

CONCLUSION

This section has presented a liberal Christian theological perspective by exploring four doctrinal questions. At the edges of the responses are other questions that are significant but not essential to an environmental ethics. Answering these four doctrinal questions shows that a liberal theological perspective for environmental ethics points coherently in a consistent direction. There is, to be sure, no absolute homogeneity within the liberal theological category, but there is sufficient coherence to make exploring the topic in this manner worthwhile.

Having examined a generic liberal approach to environmental ethics, the next section examines the theological perspective of a particular figure within the liberal tradition, namely Joseph Sittler, arguably the founder of Christian environmental ethics.

CHAPTER SIX

THE LIBERAL ENVIRONMENTALISM OF JOSEPH SITTLER

Joseph Sittler grew up in a pastoral setting. He was born in 1904 in Upper Sandusky, Ohio, an area still known for open fields and farms. During Sittler's childhood, his father was a Lutheran pastor for several rural Midwestern congregations. This allowed Sittler, at an early age, to appreciate the need for a theology of all of creation "that can penetrate the ordinary problems of human existence, including the care of the earth."[1]

Upon graduation from seminary, Sittler served as pastor of Messiah Lutheran Church in Cleveland Heights, Ohio, for thirteen years. During this time he continued his education at the University of Heidelberg, Case Western Reserve, and Oberlin Theological School. He was appointed as Professor of Systematic Theology at Chicago Lutheran Seminary in 1943, where he continued his studies at the University of Chicago, though he never completed a PhD.[2] Despite this he served as professor of theology at the University of Chicago from 1957 until his retirement in 1973.[3]

Although Sittler's *corpus* is limited—only eight slim monographs and a host of shorter articles, editorials, and devotionals in academic and

---

1. See Bakken, "Nature as a Theater of Grace," 2n4.

2. These biographical details drawn from Bakken, "Nature as a Theater of Grace," 2–3.

3. Sittler, *Grace Notes and Other Fragments*, 125.

popular periodicals—his influence has been broad. He was the president of the American Theological Society in 1951, an active participant in the theological development of American Lutherans, and a delegate to the World Council of Churches.[4] In addition to his theological work, he was a popular preacher and often cited for his pleasant demeanor, though he is described as an absent-minded professor.[5]

He is not a household name, like some other twentieth-century scholars, but Sittler warranted a *festschrift*,[6] multiple dissertations on his theology,[7] articles celebrating his impact during his life,[8] and several memorial columns in theological periodicals after his death in 1987.[9] Sittler is a theological figure that must be considered when exploring the development of contemporary environmental ethics.

Sittler's theology is liberal.[10] He is not a strict modernist who denies the supernatural, but he endorses the sort of liberalism which views Scripture as authoritative in some sense but not strictly true in all cases. Gary Dorrien describes Sittler's theology as "difficult to categorize."[11] According to Dorrien, Sittler "leaned closer to neo-orthodoxy than to Chicago liberalism."[12]

The Chicago liberalism Dorrien refers to is an individualistic, experience-based faith that largely rejects the possibility of personal knowledge of God. That version of Christian theology leads to process theology and the formulation of indistinct, philosophically-oriented expressions of religion.[13] Sittler's theology is a far cry from such abstract, academically-oriented theologies. As he writes, "My theology . . . is a theology of the

---

4. See Herhold, "Probings by Joseph Sittler," 915.

5. See Marty, "Foreword," vii–viii.

6. Hefner, *The Scope of Grace*.

7. Bakken, "The Ecology of Grace"; Bouma-Prediger, *The Greening of Theology*; Siemsen, *Constructing a North American Theology through the Work of Joseph Sittler*.

8. E.g., Brauer, "Special Issue Honoring Joseph Sittler," 97–165; Cromie, "Feminism and the Grace-Full Thought of Joseph Sittler," 406–08.

9. Hefner, "Sittler, Joseph A., 1906–1987," 82–83; Klein, "Joseph Sittler Remembered," 5–28; Marty, "Mentor to Many," 95.

10. This is also evident in his historical revisionism: Santmire, "In God's Ecology," 1301; Smith, "Toward a Lutheran Theology of Nature," 147–51.

11. Dorrien, *The Making of American Liberal Theology*, 125.

12. Dorrien, *The Making of American Liberal Theology*, 125. Sittler rejects neo-orthodoxy several times, e.g., Sittler, "A Theology for Earth," 24.

13. Dorrien, *The Making of American Liberal Theology*, 58–97.

incarnation applied to nature."[14] Still, his theology shares the rationalistic tendencies of many modern liberals.

Although Sittler recognizes Scripture as a source for theology, he often finds authority for his theological points from poetry, observations, and contemporary cultural understandings.[15] In an essay on Christology, Sittler writes, "We recognize that the classical Christology of the Creeds perpetuates formulations which operate with a way of speaking about God which is incongruent with our time and its way of thinking."[16] Rather than trying to translate the creeds to contemporary language, he tended to treat the Zeitgeist as authoritative and modify his theology to fit. Sittler formulated a theology shaped in both essence and expression by the surrounding modern culture.

The immanence of God in Sittler's theology can be seen in his cosmic Christology. Sittler understands God to be at work directly in the world through the power of Christ. For Sittler, "God simply *is* what God manifestly does."[17] The reality of God, which Sittler holds to be self-evident in the world, drives ethical action. The foundation of ethics is in re-enacting the life of Christ, the incarnate God-man.[18]

The moralization of dogma also connects to Sittler's theology. It is not the doctrinal belief *per se*, but the ethical actions which derive from it that are significant to Sittler. Indeed, Sittler declares his purpose "to articulate the immediacy of grace, to interiorize the objective reality of the dogma so that it shall become forceful for our time's need to stand *within* the creation as we receive redemption."[19] It is not the truth of nature and grace that is his primary objective, but a resulting orthopraxy.[20]

The final theme Olson highlights is universal salvation of humanity. On this point, Sittler is difficult to nail down, in part because he treats few doctrines systematically in his published works. For example, he never addresses the doctrine of atonement.[21] However, Sittler frequently

---

14. Sittler, *Gravity and Grace*, 54.
15. Heggen, "Dappled Things," 29–43.
16. Sittler, "A Christology of Function," 122.
17. Sittler, *The Structure of Christian Ethics*, 4.
18. Sittler, *The Structure of Christian Ethics*, 48. For Sittler, Christology is ultimately an eschatological category, which is why it is not treated separately in this discussion. See, Sittler, "Christ and the Moral Life," 343.
19. Sittler, *Essays on Nature and Grace*, 5. Emphasis original.
20. Sittler, "Called to Unity," 12.
21. Siemsen, *Embodied Grace*, 109.

treats sin as sickness and alienation rather than death.[22] He also rejects the notion "that God is concerned to save men's souls!"[23] This may be seen as a rejection of dualism, but seems to communicate more in context. And, at the same time, Sittler's insistence that "all things are permeable to his cosmic redemption because all things subsist in him" seems to allow for a universal reconciliation of humanity regardless of explicit faith in Christ.[24]

Sittler is a liberal theologian, though his approach to theology does not reflect the radical rejection of Christian orthodoxy of those who preceded him at the Chicago Divinity School. Sittler strives to reconfigure traditional Christian doctrines to accommodate contemporary concerns, which serves to place him within the liberal theological stream.

Identifying a cogent theological perspective in Sittler is, at times, difficult because his writings are largely occasional. Additionally, Sittler rejects theological methodology, leading to inconsistency. Under the heading "Theological Method" in *Essays on Nature and Grace*, Sittler writes, "My own disinclination to state a theological method is grounded in the strong conviction that one does not devise a method and then dig into the data; one lives with the data, lets their force, variety, and authenticity generate a sense for what Jean Daniélou calls a 'way of knowing' appropriate to the nature of the data."[25] Sittler is attesting to the flexible application of his theological perspective.

Despite this flexibility, Sittler is not unsystematic. Elaine Siemsen addresses the lack of methodology, arguing the systematic nature of Sittler "is found by understanding the term system as found in the organic experience."[26] Sittler himself defines Martin Luther's systemic vision as "a central authority, a pervasive style, a way of bringing every theme and problem under the rays of the central illumination."[27] Despite variable

---

22. For example: Sittler, *The Structure of Christian Ethics*, 57; Sittler, "Nature and Grace: Reflections on an Old Rubric," 254; Sittler, *The Care of the Earth*, 27.

23. Sittler, "A Theology for Earth," 29.

24. Sittler, "Scope of Christological Reflection," 13; Sittler, "Called to Unity," 6. In an ecumenical conference on the attitude of Christians toward Jews, Sittler demonstrate no concern for salvific differences between the religions: Sittler, "Judeo-Christian Themes," 104–08. Additionally, Rasmussen seems to read universalism in Sittler's work: See Rasmussen, "Luther and a Gospel of Earth," 9–13.

25. Sittler, *Essays on Nature and Grace*, 20.

26. Siemsen, *Embodied Grace*, 87.

27. Sittler, *The Doctrine of the Word of God*, 4.

expressions in Sittler's theological system, the following section examines his theological perspective under traditional doctrinal headings.[28]

REVELATION

Revelation is one of few doctrines Sittler addresses systematically in his published works. In a small book, *The Doctrine of the Word in the Structure of Lutheran Theology*, which is derived from a series of lectures, Sittler presents a revisionist understanding of Scripture that necessarily includes experience, reason, and tradition as theological sources roughly equal in standing to Scripture. This broad subject is addressed occasionally in his other published works as well, which helps demonstrate the consistency of his approach to the doctrine over time.

For Sittler, the word of God is not propositional in nature.[29] Although Scripture records both historical facts and logical propositions, equating the word of God with Scripture is "confusing things heavenly with things historical."[30] According to Sittler, "Historical facts in themselves do not constitute a revelation. The Spirit, who gives faith, makes historical facts revelatory. These facts become carriers of God's word because they are addressed to faith and we, before them, may be receptors of the word they carry only by faith."[31] Sittler also asserts, "Revelation is not a thing; it is a continuing activity. It is not static, but dynamic." It is because of the changing nature of revelation Sittler rejects the development of ethical principles, because such principles are "not adequate to express the new character of the Christian life."[32]

By redefining revelation to exclude a stable, comprehensible canon, Sittler opens the door to a more significant role for other sources of theology. By describing revelation as "a continuum, [where] every moment, including the present one, is a part of it,"[33] contemporary influences can redefine the whole theological schema to suit the mood of the day.

---

28. This systematization of the relatively unsystematized is consistent with the approaches used in Bouma-Prediger, *The Greening of Theology*; Siemsen, *Embodied Grace*.

    29. Sittler, *The Doctrine of the Word of God*, 1.

    30. Sittler, *The Doctrine of the Word of God*, 11.

    31. Sittler, *The Doctrine of the Word of God*, 10.

    32. Sittler, *The Doctrine of the Word of God*, 11.

    33. Sittler, *The Doctrine of the Word of God*, 61.

Sittler requires a fluid reading of Christian tradition—in which category he includes Scripture—for Christian ethics because "any effort to ground and elaborate an ethics for Christian faith must in every situation and generation shake off impatiently the investitures of ethos and habitual piety, and renew, criticize, and correct its response at this place of impact."[34] Sittler cites Jesus' Sermon on the Mount as an indication of the impossibility of deriving ethical principles from true revelation, noting the absence of particular, practical mandates.[35] The authority of Scripture is found in its ability to connect with the human condition and inspire, not in its content.[36]

This approach frees the contemporary ethicist from the tyranny of potentially flawed traditional interpretations, but it also undermines the role of Scripture within the realm of Christian ethics. Instead of an available, consistent source of God's voice for moral decisions, Sittler proposes an interpretation of God's ongoing revelation in the world.[37] This form of moral theology, "in which faith is the determinant in all things from the doctrine of God to the doctrine of the good (ethics), arises from this controlling fact: that all gifts from God to man are given in and pass through the historical; and the historical, *qua historical*, can never beget life, certainty, [or] redemption."[38] Sittler rightly recognizes the morass of moral uncertainty such an approach places humans in, but argues when faith is the central tenet of the God-human relationship, "then the actual situation of decision-pressed man is saved from the despair which would inevitably overtake him" if the relationship were founded on mere love.[39]

Sittler's belief in ongoing revelation apart from Scripture is closely tied to his high view of the undistortedness of creation and the *imago Dei* even despite the fall.[40] The impact of sin on the created order is of such little concern that he calls for literature outside of Scripture to be used as authentic revelation of God to drive ethical choices.[41] Sittler writes, "As theologian and preacher, literature has never been for me merely a mine

---

34. Sittler, "Ethics and the New Testament Style," 30.
35. Sittler, "Ethics and the New Testament Style," 30.
36. Sittler, *Gravity and Grace*, 34, 45.
37. Sittler, *Gravity and Grace*, 51.
38. Sittler, "Ethics and the New Testament Style," 34.
39. Sittler, "Ethics and the New Testament Style," 35. From the context and timing, it seems likely Sittler is responding to Fletcher, *Situation Ethics*.
40. Siemsen, *Embodied Grace*, 120.
41. Sittler, *The Structure of Christian Ethics*, 17–20.

of felicitous articulation of things that have been elsewhere learned; it has itself been a place of listening and learning."[42] This is why Sittler frequently cites literature and poetry as authoritative sources for his theology, seemingly equal to Scripture.[43]

Since Scripture is insufficient for developing a Christian ethics, Sittler roots his ethics in the contemporary situation.[44] This, in turn, redirects the fundamental aim of Christian ethics from obedience to God's moral norms to creating "faithful re-enactments of [Christ's] life."[45] According to Sittler, this breaks down wooden legalism "and out of the living, perceptive, restorative passion of faith enfolds in its embrace the fluctuant, incalculable, novel emergents of human life."[46] Sittler escapes legalism, but like others in the Lutheran tradition ends with a form of antinomianism that is detrimental to obedience.

The derivation of ethical norms from contemporary context with the aim of living like Christ sounds spiritual. However, such an approach does very little to help faithful Christians live consistently in their context. Sittler tells Christians to act like Jesus, but he undermines reliability of the only source that describes Jesus' nature.[47] The rejection of Scripture as the supreme source of revelation marks the creation of an impossible Christian ethics that necessarily devolves into individual conscience and personal convictions.

CREATION

It should go without saying that the doctrine of creation is central to an expressed environmental ethics. For Sittler this has less to do with cosmological questions and the age of the earth than with how value is assigned to nature.[48] Sittler's approach to the question of value in nature is theocentric. Creation has value because of the one who created it, and he

---

42. Sittler, *Essays on Nature and Grace*, 19.

43. E.g., Sittler, "In the Light of Our Biblical Tradition," 191–92; Sittler, *Grace Notes and Other Fragments*, 119; Sittler, "Nature and Grace in Romans 8," 217.

44. Sittler, *Gravity and Grace*, 57.

45. Sittler, *The Structure of Christian Ethics*.

46. Sittler, *The Structure of Christian Ethics*, 48.

47. He writes, "We have no access to Jesus"; Sittler, *Gravity and Grace*, 13.

48. Sittler argues Genesis does not address the age of the earth. Sittler, *The Structure of Christian Ethics*, 5.

has remained immanent in the world. He writes, "God is the Creator. He is the fountain of life from whose eternal livingness all things are brought forth . . . God is not identified with the world, for he *made* it; but God is not separate from his world either. For *he* made it."[49]

Value is rooted not in the creation itself, but in the one who created all things. Sittler insists, "Because God is Creator his reality is not to be denied by idolatrous substitution of any earthly source, good, value, or purpose as ultimate."[50] Sittler seems to be illustrating this when he invokes categories for *use, enjoyment,* and *abuse* in "The Care of the Earth." Sittler paraphrases Aquinas's interpretation of these categories: "It is the heart of sin that men use what they ought to enjoy and enjoy what they ought to use."[51] *Enjoyment* in this case is the *use* of a part of the created order as a gift from God in proper relatedness to the Creator. This gracious acceptance and *use* of the material world is consistent with Sittler's description of the nature of a Christian: "To be a Christian is to accept what God gives."[52]

In Sittler's mind, one of the chief problems with modern theology is alienation of humans from nature.[53] Thus he develops a creationally grounded Christology designed to highlight human relatedness with creation.[54] This, not propitiation of sins, is the primary purpose of Christ's incarnation. Sittler's "cosmic Christology" is really a "creational Christology" that emphasizes God's immanence in the created order, rather than emphasizing his condescension to redeem creation from the curse.

For Sittler, creation is in need of redemption, but redemption is merely restoring proper relations between God-man-nature.[55] Christ's life as God-man demonstrates the restoration of the right relationship between God and nature, as his miracles are enacted parables designed to represent the redeemed creation.[56] Thus Sittler describes the essence of Christian ethics as "faithful re-enactments of [Christ's] life."[57]

---

49. Sittler, *The Structure of Christian Ethics*, 5.
50. Sittler, *The Structure of Christian Ethics*, 70.
51. Sittler, "The Care of the Earth," 56.
52. Sittler, *The Structure of Christian Ethics*, 87.
53. Sittler, "A Theology for Earth," 29.
54. Sittler, *The Structure of Christian Ethics*, 7–8.
55. Sittler, "Nature and Grace in Romans 8," 209–11.
56. Sittler, *The Structure of Christian Ethics*, 51; Sittler, *Gravity and Grace*, 58–59.
57. Sittler, *The Structure of Christian Ethics*, 48.

At times, Sittler's language about the created order is clouded.[58] He functions within a Lutheran framework, frequently seeking to find points of contact between his worldview and Luther's.[59] Luther's two kingdoms approach assumes a radical disruption of the created order, seeing a division between nature and grace which requires God's action to restore.[60] But Sittler sees no such division, arguing that a central role of Christians is to rightly relate nature to grace, though he seems to be arguing for a reformulation of human thinking rather than redemptive action.[61]

Sittler's *telos* for creation is to passively allow the human mind to relate to an immanent God rather than to actively glorify a transcendent God.[62] The human who is seeking God can understand him through creation. Sittler resists pantheistic identification of God with creation, but he diminishes the Creator-creature distinction.[63] In attempting to reunite humans with the created order, Sittler diminishes God's otherness. He establishes an ethics of *being* that denies full biblical purpose for the created order. As a result, in his anthropology, humans are left without a distinct purpose.

ANTHROPOLOGY

Sittler's understanding of anthropology connects to his doctrine of creation through his relational ethics. In an essay outlining the nature of man, Sittler writes, "He is *relational* in his structure; and his ultimate relationship is a Creation. This relationship not only establishes his ontology; it also determines his history as a person, and gives a strange dialectic to his evaluation of human history."[64] This relationship between humans and creation is essentially unification of humanity with creation.

Sittler attempts to establish a theology for the earth that restores balanced relationships in the God-man-nature triad. Humans are to be concerned with rightly relating to the created order. Thus humanity's

---

58. Siemsen, *Embodied Grace*, 18–20.

59. See particularly his attempt to reconcile his revisionist approach to Scripture to Luther's high view of Scripture: Sittler, *The Doctrine of the Word of God*, 13–32.

60. Goheen and Bartholomew, *Living at the Crossroads*, 61–62.

61. Sittler, "Commencement Address," 36.

62. Sittler, *The Structure of Christian Ethics*, 7–8.

63. Sittler, *The Structure of Christian Ethics*, 4.

64. Sittler, "In the Light of Our Biblical Tradition," 188.

ethical duty is to rightly receive what God gives. Sittler writes, "The Christian man is to accept what God gives as Creator: the world with its needs, problems, possibilities; its given orders of family, community, state, economy. Each of these is invested with the promise and potency of grace, and each of these is malleable to the perverse purposes of evil."[65] At one level, this reflects a communitarian approach to ethics, by seeking right relationships at all levels. However, at the same time it makes Christian ethics a largely individualistic enterprise, where everything is dependent upon personal response to God's gift.

There is promise in Sittler's approach, but such ambiguity simultaneously diminishes hope for praiseworthiness. This is, in part, because a strong existential strain runs through Sittler's description of biblical ethics when he writes, "Men are not called to an ideal, or threatened with failure to match an elevated standard of abstract goodness. They are called rather to be what they are, live their true life, realize their being in their existence, and work out their relationships on earth in organic continuity with their relationship to the Creator."[66] Rightly relating to the created order is essential for relating properly to God because Sittler finds an "organic continuity between God and nature."[67] Sittler's ethics are earthy, which is a helpful corrective to the sometimes dualistic perspective of popular Christianity. However, it seems to limit the ability of humans to directly relate to God.

As Sittler understands it, humanity "lives in a nexus of human relations and nature relations in which sheer operational facticity pierces into and deeply complicates all efforts to discern, clarify, and respond to the righteousness of God."[68] The duty of humans is to rightly relate to God and creation, but that duty is frustrated by the human situation within creation. Thus Sittler despairs of establishing "an ethics that shall have the coherent authority of older systems."[69] This leads to his understanding that "there is, to be sure, no human fact in which sin is not involved."[70] This formulation increases the human need for divine grace, but it also leaves humans in a quagmire of ethical reasoning with little

---

65. Sittler, *The Structure of Christian Ethics*, 87.
66. Sittler, *The Structure of Christian Ethics*, 6.
67. Sittler, *The Structure of Christian Ethics*, 7.
68. Sittler, *Essays on Nature and Grace*, 17.
69. Sittler, *Essays on Nature and Grace*, 18.
70. Sittler, *The Structure of Christian Ethics*, 86.

hope of moral clarity. The answer for Sittler is faith that keeps people from despair.[71] In the end, the lack of clearly available special revelation from God leads to an inability to determine right from wrong and results in Sittler's appeal to an ambiguous notion of faith, which is really an appeal to mystery.[72] This is an epistemic problem that Sittler's anthropology never overcomes.

Despite this weakness in his anthropology, Sittler is no misanthrope. He does not oppose human population and technological development. In fact, Sittler sees cities "as fair a field for the promises of God, as ever the rural situation was."[73] Sittler, therefore, does not eschew culture, though he recognizes the terrifying power humans can exert over creation: "[Man] is in the world as a creature who, for the first time, has the structures and powers of the natural world sufficiently and growingly within his manipulative grasp to form it toward an order for human life of dreamlike potentialities, or to utilize his knowledge-power toward an unimaginable hell."[74]

In line with traditional interpretations of the human role in creation, Sittler owns there is a special responsibility for humans in the created order to "tend as God's other creation."[75] Thus humans have a redemptive role on earth. Indeed, in this, humans are participating in Christ's cosmic work of redemption. According to Sittler, "Our task, as I understand it, is to restore the right relationship of nature and grace—in obedience, in word and in thought."[76]

Sittler's anthropology reinforces the need for developing an environmental ethics. His recognition of humans as created, but distinct from the created order, aids in inculcating a sense of responsibility for caring for the environment. Likewise, Sittler's promotion of the relatedness of God, humans, and creation benefits his environmental ethics. Moreover, his acceptance of the possibility of technological advancement consonant with proper environmental stewardship is more consistent with a biblical notion of ecological responsibility than some contemporary environmentalists allow. In all, Sittler's anthropology is the most helpful aspect

---

71. Sittler, *The Structure of Christian Ethics*, 83.
72. Sittler, *The Structure of Christian Ethics*, 83–85.
73. Sittler, "Commencement Address," 36.
74. Sittler, *Essays on Nature and Grace*, 17.
75. Sittler, "A Theology for Earth," 29.
76. Sittler, "Commencement Address," 36.

of his theological perspective in developing a biblical environmental ethics. At the same time, the epistemic barrier he erects between humans and special revelation significantly tempers any positive contribution his anthropology makes to an overall theological schema.

ESCHATOLOGY

Eschatology is present in Sittler's theology, for it is tied up with his doctrines of revelation, creation, and anthropology. It is, however, a diminished version of eschatology compared to other theological perspectives. The focus of Sittler's eschatology is the limitedness of humanity within the God-man-nature relationship, as he notes, "This relationship not only establishes his ontology; it also determines his history as a person, and gives a strange dialectic to his evaluation of human history."[77] It is the mysterious future, which is to be revealed later, that provides a sense of ethical duty after Sittler has abandoned ethical absolutes when he states, "At one and the same time [eschatology] relativizes all things human and historical, and makes the human decision and the historical behavior absolutely crucial."[78]

In a volume on preaching, Sittler deals with eschatology under the heading, "The Tyranny of Boundlessness."[79] It is into this boundlessness, which is characteristic of traditional Christian theology, that "the eschatological and bounded finalities of the gospel are absorbed without great effect."[80] In contrast, the function of eschatology is in part to show the magnitude of God, but also to emphasize the proper limits of the human experience.[81] Eschatology, for Sittler, is an anthropocentric doctrine that is most helpful when it limits the expansive sense of possibility.[82]

Sittler's Christianity is global, as evidenced by his work with the World Council of Churches, but it is nevertheless deeply embedded in his North American context. Speaking to that context, Sittler critiques expansive eschatologies that looked hopefully toward the future as a product of the early American psyche, which was fueled by the seeming

---

77. Sittler, "In the Light of Our Biblical Tradition," 188.
78. Sittler, "In the Light of Our Biblical Tradition," 189.
79. Sittler, *The Ecology of Faith*, 14–25.
80. Sittler, *The Ecology of Faith*, 15.
81. Sittler, *The Ecology of Faith*, 23–25.
82. Siemsen, *Embodied Grace*, 28–29.

limitlessness of the frontier.[83] This expansive hope, according to Sittler, led earlier Americans to believe they could continue to relocate away from problems, that there would continue to be new ecological possibilities. This common memory leads contemporary Americans to neglect of the environment. He writes, "The accumulated garbage of the achievement [of society and industrialization] has befouled the air, polluted the water, scarred the land, besmirched the beautiful, clogged and confused our living space, so managed all placement and means of movement as to convenience us as consumers and insult us as persons."[84] One purpose of eschatology, therefore, is to rein in the sense of limitlessness and force humans to accept their finiteness. It makes human decisions in space and time significant because there are limits preventing avoidance of consequences. However, Sittler's emphasis on the limitedness of humans through eschatology also tends to undermine the accessibility of ethical norms.

A second purpose of eschatology, tied up with the concept of humanity's limitedness, is to reinforce unknowability of ethical absolutes while preventing despair. Sittler's limiting eschatology is supposed to characterize the way Christians lived in the first century. He writes, "This eschatalogical mode of existence alone explains what were otherwise incomprehensible: an *active* passivity, ethically absolute commands delivered to relative capacities, discouragement borne without despair, men cast down but not destroyed."[85] This mode of existence is intended to continue into the contemporary era.

Despite its optimism, Sittler's eschatology does not take into account the power of God to work within contemporary Christians to reveal absolutes, thus undermining their existence as new creations.[86] Sittler's eschatology is over-realized, describing nature and grace as essentially unified in the present.[87] Critics of Sittler's cosmic Christology note that he calls Christians to look backward toward creation instead of forward toward the coming kingdom of God.[88]

---

83. Sittler, "Space and Time in American Religious Experience," 45–47.
84. Sittler, "Space and Time in American Religious Experience," 50.
85. Sittler, "Dogma and Doxa," 10.
86. Sittler, "An Open Letter by Joseph Sittler," 270.
87. Bouma-Prediger, *The Greening of Theology*, 75–85; Sittler, *Essays on Nature and Grace*, 75.
88. Siemsen, *Embodied Grace*, 116.

Personal eschatology is a topic on which Sittler evolved over his career. In his 1958 treatise, *The Structure of Christian Ethics*, Sittler appears to anticipate a literal, future realization of cosmic redemption.[89] In 1975, in a sermon on Romans 8, Sittler argues, "The Christian hope of eternal life... is not an unreasonable one."[90] Later in life, Sittler openly questions the possibility of a future bodily resurrection.[91] He declares heaven to be merely "a metaphor for the fulfillment of life in God,"[92] and he adamantly proclaims, "The Bible really has nothing to say about eternal life."[93]

At times in Sittler's theology, there seems to be an expectation for human participation in the redeemed New Creation in the future, but it is not a reality Sittler seems to believe will really happen.[94] Or, perhaps it might be better said that the future reality may be an improvement in the condition of creation that might, through incremental changes, one day be called the kingdom of God. It is not clear whether such a future state will include individual humans actually living in the presence of God for all eternity.[95]

Sittler's eschatology is redemptive for creation, seeing a future fulfillment of the potentials of the created order.[96] However, by expanding the place for creation as a recipient of grace, Sittler seems to have diminished the place of individual humans within the redemptive power of grace.[97] This has significant implications for his environmental ethics.

The most significant impact of eschatology on Sittler's environmental ethics is the rejection of dualism. The eschatological theme of divine redemption of the created order leads to a strong sense of duty to participate in Christlike fashion in God's redemptive work. Sittler's version of eschatology also limits the role of humanity in creation, inculcating a sense of boundedness for Christian ethics. This limiting prevents humans from seeing themselves as primary actors in redemption or as tyrannical

---

89. Sittler, *The Structure of Christian Ethics*, 51–52.

90. Sittler, "Nature and Grace in Romans 8," 217.

91. Sittler, *Grace Notes and Other Fragments*, 119–21. His longer essay is more emphatic: Sittler, "The Last Lecture," 59–65.

92. Sittler, *Gravity and Grace*, 73.

93. Sittler, *Gravity and Grace*, 74.

94. Sittler, *Essays on Nature and Grace*, 119–20.

95. Sittler, *Essays on Nature and Grace*, 109–11.

96. Sittler, *The Structure of Christian Ethics*.

97. Sittler, *Essays on Nature and Grace*, 106–11.

overlords of creation with little or no accountability for stewardship of the cosmos.

At the same time, Sittler's eschatology diminishes the hope of actionable ethics. For Sittler, eschatology reinforces the conditional, unknowable nature of ethics. He seems to believe this will improve human attitudes toward the environment, but at the same time, by his understanding, eschatology also serves to undermine absolutes within ethics. The uncertainty of the ethical quality of actions and even the future state of creation provides a reasonable basis for present passivity or creative redefinition of norms, which are as likely to have detrimental consequences for creation as redemptive. Thus he lays groundwork for ethics in a different era to have more or less concern for creation depending on the influences of the day.

CONCLUSION

Sittler's approach is a prime example of theological reformulation to meet perceived ethical needs.[98] In other words, based on his answers to the four questions in his theological perspective for environmental ethics, his theological foundations—not just his ethical outcomes—have been modified. Such an approach contrasts with a more theologically conservative approach which plumbs the depths of Christian theological sources to formulate cogent ethical responses to contemporary crises, but which still maintains the essential forms of earlier Christian tradition.

Sittler promoted a theology *of* ecology as opposed to a theology *for* ecology as he maintains, "For if we start talking about a theology *for* ecology, we will try to manufacture out of the uncriticized theological categories consequent moralistic efforts stretched to enclose new and crucial facts. Such an effort will not really be a redoing of theology in view of ecology but only an extension of traditional ethics in the presence of crisis."[99] Indeed, Sittler argues, "In every field of science and technology, ecological fact is presently so clear and the result of its ignoring so catastrophic that older guides to action are either useless or positively perilous."[100] Hence, reformulation of dogmatic expressions to meet con-

---

98. Sittler, "An Open Letter by Joseph Sittler," 270–72.

99. Sittler, "Ecological Commitment as Theological Responsibility," 35. Emphasis original.

100. Sittler, *Essays on Nature and Grace*, 112.

temporary needs is an ethical imperative based on pressing social conditions. This process can be seen at work in the four doctrines discussed in this project.

Sittler creates a theology, and thereby an ethics, which addresses the pressing issues of his day.[101] There is a relative absence of the background explaining exactly what the particular issues of the day were to which Sittler was responding. When Sittler does mention particular issues that require response, it is with a vague assumption of the self-evidence. In one address he states, "There obviously is emerging a politics of theology. There is already a well-developed statistics of ecology."[102] While these assertions are even more evident in the twenty-first century, the list of concerns has shifted dramatically, which may bring into question how well Sittler's theological perspective can serve contemporary environmental ethics. Nevertheless, if one assumes that the main concern is detrimental human impact on the environment, then Sittler's theological response can be fairly evaluated in the present milieu.

There are few clear ethical mandates within Sittler's environmental ethics. This is in part because of his understanding of revelation and eschatology. After certainty in Scripture is removed and human limitation in eschatology accepted, the central goal of his ethics becomes a restoration of the "right" relationship between humans and creation. In this restored order, "Man and nature live out their distinct but related lives in a complex that recalls the divine intention as that intention is symbolically related on the first page of the Bible. Man is placed, you will recall, in the garden of earth. This garden he is to tend as God's other creation—not to use as a godless warehouse or to rape as a tyrant."[103] Sittler rejects propositions within Christian ethics, looking for a continuously adaptable, situationally flexible ethical approach.[104] He writes,

> A Christian ethics must, therefore, work where love reveals need. It must do this work in faith which comes from God and not as accumulating achievements to present to God. In this working it must seek limited objectives without apology, and

---

101. Sittler, *Essays on Nature and Grace*, 117.

102. Sittler, "Ecological Commitment as Theological Responsibility," 35; Sittler, "Called to Unity," 10.

103. Sittler, "A Theology for Earth," 28–29.

104. Sittler, "In the Light of Our Biblical Tradition," 189.

support failure without despair. It can accept ambiguity without
lassitude, and seek justice without identifying justice and love.[105]

Thus, Christian environmental ethics are "the actualization of justification" to all of creation.[106] This is not the sort of environmental ethics that leads to easy application in the form of a how-to manual, but it is the framework Sittler provides. It is intrinsically theological and intertwined with his doctrines of revelation, creation, anthropology, and eschatology.

Though other doctrines could be used to enrich this matrix of influence of theological perspective over Sittler's environmental ethics, the four doctrines discussed here are sufficient to illustrate the way Sittler's theology shapes his environmental ethics. To understand Sittler's environmental ethics, one must understand his answers to these four questions.

Sittler's doctrine of revelation makes Scripture largely inaccessible to the modern mind, forcing Christians to seek ethical norms only from their understanding of the life of Jesus and current data. This makes his ethics adaptable to modern narratives and cultural concerns. His doctrine of creation emphasizes the inherent goodness of creation; there is value in creation because it is, and always has been, rightly related to God, except where humans have interrupted the relationship. Anthropology structures Sittler's environmental ethics by placing humans within the web of the created order, but distinct. Thus humans have the responsibility to act with something more than self-interest to restore the God-man-creation relationship to its intended state. Finally, Sittler's eschatology creates a limited sense of human possibility in the world and emphasizes the ambiguity of ethical knowledge. For Sittler, eschatology is not primarily a doctrine of the future state but of the present limited state of humanity. It is a doctrine of humility.

Sittler's environmental ethics are distinctly theological. His ethical schema is based on a liberal reformulation of Lutheran theological foundations applied in light of concern for environmental degradation. Thus his theological perspective controls the formulation of his environmental ethics.

---

105. Sittler, *The Structure of Christian Ethics*, 84.
106. Sittler, *The Structure of Christian Ethics*, 72.

PART FOUR

A FUNDAMENTALIST
ENVIRONMENTAL
ETHICS

CHAPTER SEVEN

A Fundamentalist Perspective for Environmental Ethics

THE FUNDAMENTALIST CHRISTIAN THEOLOGICAL perspective for environmental ethics is more difficult to discuss than the other three categories in this project because of a relative dearth of source material. Very few fundamentalist Christians articulate extended treatments of environmental ethics.[1] This gnomon is the result of several factors. First, fundamentalists tend to be less active in producing academic treatments of any topic than Christians in other theological streams. Second, fundamentalists have largely resisted engagement with environmental ethics because they perceive it to be an issue of secondary concern, not central to the gospel. Third, some fundamentalists consider environmental ethics a "liberal issue;" thus, the fundamentalist tendency toward separation has limited their engagement in the discussion.

In light of the dearth of fundamentalist Christians writing positively about environmental ethics, an attempt to diagnose a fundamentalist theological perspective for environmental ethics warrants justification. One significant reason to focus on the explication of a fundamentalist position on environmental ethics (as much as one exists) is to dispel the

---

1. One possible exception is Joel Salatin, *The Marvelous Pigness of Pigs*. Salatin is a Presbyterian who graduated from Bob Jones University and has since been honored as an alumnus of the year by BJU.

common confusion that many environmentalists seem to have about the differences between fundamentalists and evangelicals. For example, Katherine Wilkinson, in her history of evangelical attitudes toward climate change, includes fundamentalists within the category of evangelicals.[2] Additionally, many environmentalists describe E. Calvin Beisner as the representative of the fundamentalist perspective on environmental ethics.[3] Beisner affirms the fundamentals of the faith, and is thus, in one sense, a fundamentalist. However, his willingness to cooperate with others of varying theological persuasions and degrees of separation from error rules him out as a fundamentalist as it is defined for this project. In the case of Beisner and his organization, the Cornwall Alliance, it seems more likely that those who use the "fundamentalist" label are attempting to dismiss Beisner's perspective to avoid grappling with the theological concepts he advocates.

In light of the need for disambiguation, this chapter offers a definition of "fundamentalism." Then, it presents critiques of a fundamentalist environmental ethic, which help to show what the common understanding of the position is. Next, it uses several recognized fundamentalist theologians to demonstrate a "typical" approach to the four doctrines in question: revelation, creation, anthropology, and eschatology, with a view to finding answers to the four basic questions of a theological perspective on the environment. In addition to demonstrating the adequacy of a theological perspective for environmental ethics, this chapter shows that it is possible to construct a fundamentalist environmental ethic based on their existing systematic theologies.

DEFINING FUNDAMENTALISM

Of the theological types discussed in this volume, I believe the most misunderstood position is that of the fundamentalist. Few labels can do more damage to a person's reputation than "fundamentalist." As self-described

---

2. Katharine K. Wilkinson, *Between God and Green*, 5.

3. At conferences, I frequently query colleagues writing on environmentalism about fundamentalist sources and am nearly always directed to Beisner's work, especially his work in support of the book and related film series that openly opposes the environmental movement. Wanliss, *Resisting the Green Dragon*. Other sources, like a multi-authored assessment of different streams of hermeneutics all relating to environmental ethics, are more explicit. Horrell, Hunt, and Southgate, "Appeals to the Bible in Ecotheology and Environmental Ethics," 219–38.

fundamentalist Kevin Bauder notes, "As a term of theological opprobrium, *fundamentalist* is about as bad as it gets."[4] The epithet brings to mind crazed terrorists flying planes into towers, angry mobs burning books, and constricting social practices. To some, the idea of a fundamentalist is the same as a bigot as it is properly defined—narrow-minded and mentally inflexible.[5] Philosopher Alvin Plantinga notes that for many critics "the full meaning of the term ['fundamentalist'], therefore (in this use) can be given by something like 'stupid sumbitch whose theological opinions are considerably to the right of mine.'"[6] Supporting this definition, Bauder comments, "Fundamentalism is generally treated like the cryptozoology of the theological world. It need not be argued against. It can simply be dismissed."[7] There is, however, a historical tradition of fundamentalism that deserves careful attention.

As Roger Olson writes, "The term 'fundamentalist' arouses many thoughts and feelings in people, but it was first used to name a movement of conservative Protestants in Britain and America to oppose liberal theology . . . In the early twentieth century fundamentalism was the defense of basic Christian doctrines from the acids of modernity and liberal thought."[8] Fundamentalism, by this description, is more than a cult of backward thinkers who reject theological advancement. Fundamentalists have a unique theological vision, with methods, boundaries, and concerns that warrant careful consideration.

One critic, liberal historian, Roger Gottlieb, describes a fundamentalist as someone who does not understand that he chooses his beliefs arbitrarily, just like everyone else. Gottlieb argues, "Fundamentalism arises

4. Bauder, "A Fundamentalist Response to 'Confessional Evangelicalism,'" 97. Emphasis original.

5. For example, this definition of fundamentalism from Cole gives evidence to a strong bias against the movement: "Fundamentalism was the organized determination of conservative churchmen to continue the imperialistic culture of historic Protestantism within an inhospitable civilization dominated by secular interests and a progressive Christian idealism. The fundamentalist was opposed to social change, particularly such change as threatened the standards of his faith and his status in ecclesiastical circles. As a Christian, he insisted upon the preservation of such evangelical values as at one time had been accepted universally, but in recent years were widely abandoned for more meaningful ideals." Cole, *The History of Fundamentalism*, 53. With such uncharitable criticism coming from self-described liberals, it is not hard to understand the fundamentalist withdrawal from much scholarly activity.

6. Alvin Plantinga, *Knowledge and Christian Belief*, 55.

7. Bauder, "Fundamentalism," 11.

8. Olson, *The Journey of Modern Theology*, 215.

when people are threatened by dramatic and seemingly uncontrollable change. Undermined by secularism, women's rights, technology, consumerism, and increased encounters with people of different cultures, the fundamentalist cannot accept a world in which all traditions, including his own, have become a matter of choice."[9] For Gottlieb, fundamentalism is an epistemological position that sounds something like foundationalism. The fundamentalist has an unbending mind that cannot deal with encounters with new ideas. It seems that for Gottlieb, the existence of a plurality of religions should cause the fundamentalist to question the truthfulness of his own religion.

Gottlieb attributes the fundamentalists' pursuit of their view of the proper order of society to a quest for political power, not a principled, theological perspective. As Gottlieb writes, "Facing forces that diminish his meaning, status, and social power, the fundamentalist grasps at a vision of an eternally fixed and universally true source of authority to stem the tide."[10] Based on this assessment, the societal conflict between religious conservatives and social revolutionaries is not doctrinal; it is a struggle over power.[11] When viewed through this Nietzchean lens, the

---

9. Gottlieb, *A Greener Faith*, 221. In a watershed book in moral psychology, Jonathan Haidt explores the foundations of both liberal and conservative ideologies. He finds that one of the hallmarks of conservative thought (and thereby religious fundamentalism by default) is the acceptance of authority as a main foundation of morality. This helps to explain the foundationalist roots of fundamentalism's religious response to environmentalism, especially regarding the respect for the authority of Scripture. Haidt, *The Righteous Mind*, 356–58.

10. Gottlieb, *A Greener Faith*, 221.

11. This view of fundamentalism is central to *Fundamentalism Project*, which is a series of monographs edited by Martin E. Marty and R. Scott Appleby. They write, "Religious fundamentalism has appeared in the twentieth century as a tendency, a habit of mind, found within religious communities and paradigmatically embodied in certain representative individuals and movements. It manifests itself as a strategy, or set of strategies, by which beleaguered believers attempt to preserve their distinct identity as a people or group. Feeling this identity to be at risk in the contemporary era, these believers fortify it by a selective retrieval of doctrines, beliefs, and practices from a sacred past. These retrieved 'fundamentals' are refined, modified, and sanctioned in a spirit of pragmatism: they are to serve as a bulwark against the encroachment of outsiders who threaten to draw the believers into a syncretistic, areligious, or irreligious cultural milieu. Moreover, fundamentalists present the retrieved fundamentals alongside unprecedented claims and doctrinal innovations. These innovations and supporting doctrines lend the retrieved and updated fundamentals an urgency and charismatic intensity reminiscent of the religious experiences that originally forged communal identity." Marty and Appleby, eds., *Fundamentalisms and the State*, 3. This sort of definition is typical of that offered by outsiders, and it reveals a significant bias

vitriol spewed in the public square becomes explicable. If the chief desire of the political pundit is merely to force the populace to accept a new set of traditions—which are viewed as entirely a matter of choice—then bludgeoning the opposition into submission through social ostracism, shaming, and sometimes harassment becomes not only acceptable but perhaps the preferred means of argumentation.

Gottlieb's conflation of foundationalist epistemology with fundamentalist doctrine weakens his argument, because it broadens the faction against which he is warring to include the very scientists whose data he is reporting as a support for environmental activism.[12] The chemist does not walk into the laboratory to choose whether she believes in the outcome of her experiment; rather, she goes into the lab to mix and heat the chemicals to determine what is true. The scientist's knowledge is, or so she hopes, founded in the belief that she is searching for a realizable truth.

In view of stilted definitions offered by scholars like Roger Gottleib, there is a need to make a clearer description of fundamentalism. According to Bauder, "Fundamentalism is a serious attempt to wrestle with the nature of the church as the communion of the saints."[13] This means that, even more than the definition of evangelicalism offered in this project, fundamentalism is doctrinal. The gospel itself, which is at the heart of the identity of the church, includes belief in certain central doctrines, which are essential.[14] Thus, Bauder writes, "To trust Christ as Savior is to trust a doctrinal Christ. To reject the doctrines is tantamount to rejecting Christ himself."[15] At the same time, he adds, "This does not mean that someone must explicitly know and affirm all of the fundamentals to be saved . . . No fundamental can be denied, however, without implicitly denying the gospel itself."[16]

---

against fundamentalism, which may border on projection. Such a biased definition helps explain why a psychologist and former "fundamentalist" finds it acceptable to draw close parallels between mental illness and religious fundamentalism, though he is careful to deny a direct correlation. Mercer, *Slaves to Faith*, 131–35.

12. The argument against foundationalism is also seen in Whitney Bauman's use of queer theory. Bauman, *Religion and Ecology*. Also, see Fowler comment about the general foundationalism of all Protestant environmentalism. Fowler, *The Greening of Protestant Thought*, 5.

13. Bauder, "Fundamentalism," 21.

14. Bauder, "Fundamentalism," 29.

15. Bauder, "Fundamentalism," 29.

16. Bauder, "Fundamentalism," 29.

The purity of the church is at stake if individuals with wrong doctrine are admitted as a part of it. Thus, the fundamentalist position is defined not only by belief in central doctrines, but also by the separation from those who do not embrace those doctrines. According to Rolland McCune, President and Professor of Systematic Theology at Detroit Baptist Seminary and self-described fundamentalist, the fundamentalist movement has a distinct shape apart from evangelicalism. He writes,

> [Fundamentalism] has moved in a certain direction, i.e., in a biblically conservative direction whose distinctive path has by now been well documented. Its common basis is a set of biblical doctrines and beliefs, and its motivating *esprit* is essentially its militant separatism. Perhaps other characteristics of its *esprit* such as evangelism, revival, prayer, missions, or holiness could be mentioned, but these are not really the private property of fundamentalism's defining motivation.[17]

Similarly, Bauder notes, "Fundamentalists believe that separation from apostates is essential to the integrity of the gospel."[18]

This attitude of separation has led to a relative ghettoization of fundamentalist academics.[19] McCune notes that evidence of self-definition of fundamentalists is difficult to come by because the debate is mainly conducted through "sermons, talks, or private conversations."[20] According to Bauder, early twentieth century fundamentalism is characterized, in its populist forms, by anti-intellectualism. However, this has been changing in recent years with fundamentalist seminaries and universities seeking accreditation.[21] Bauder writes, "The better fundamentalist seminaries still lag behind evangelical schools in publication, but the level of their classroom instruction is comparable."[22] The problem with engaging fundamentalists academically is a lack of documentation, which Bauder

17. McCune, "The Self-Identity of Fundamentalism," 17.

18. Bauder, "Fundamentalism," 40.

19. Another possible contributing factor to this observable phenomenon is the implicit egalitarianism built into the hermeneutics of fundamentalism. The plain-sense reading of Scripture by the common person and the rise of fundamentalism outside of academic settings led to a relative rejection of academic pursuits, or at least a view that non-academic study of Scripture was sufficient. Marsden, *Fundamentalism and American Culture*, 57–62.

20. McCune, "The Self-Identity of Fundamentalism," 13.

21. Recently Bob Jones University obtained accreditation from SACSCOC, which is a significant change in their rigid separatism from several decades ago.

22. Bauder, "Fundamentalism," 46.

self-identifies, explaining, "For a generation or more, [fundamentalists] have produced no sustained expositions of their ideas. Perhaps a certain amount of stereotyping is excusable, and maybe even unavoidable. No fundamentalist has produced a critical history of fundamentalism. Nor is any sustained, scholarly, theological explanation of core fundamentalist ideas available."[23] Not only does the dearth of fundamentalist publications lead to mischaracterization and stereotype, but it makes a project on an issue like environmental ethics, which is not at the center of fundamentalist religious belief, very difficult to document. In truth, however, the lack of documentation is not only due to a lack of publication, but also strongly related to the reluctance of many fundamentalists to see social evils as the proper field of action for the church.[24]

Carl F. H. Henry's slim volume, *The Uneasy Conscience of Modern Fundamentalism*, deals with just this issue.[25] He writes,

> The social reform movements dedicated to the elimination of such evils [e.g., aggressive warfare, racial hatred and intolerance, etc.] do not have the active, let alone vigorous, cooperation of large segments of evangelical Christianity. In fact, fundamentalist churches increasingly have repudiated the very movements whose most energetic efforts have gone into attack on such social ills. The studied fundamentalist avoidance of, and bitter criticism of, the World Council of Churches and the Federal Council of Churches of Christ in America is a pertinent example.[26]

He goes on to argue, "The great majority of fundamentalist clergymen, during the past generation of world disintegration, became increasingly less vocal about social evils. It was unusual to find a conservative preacher occupied at length with world ills."[27] This led to the "widespread notion that indifference to world evils is essential to Fundamentalism."[28] It is this

---

23. Bauder, "Fundamentalism," 19.

24. This trend has resurfaced in some conservative evangelical circles, largely in response to the concerns over the loss of the doctrinal core of the gospel in light of expansive application of the gospel to extra-ecclesial social issues. For example, see the argument in DeYoung and Gilbert, *What Is the Mission of the Church?*

25. Note that Henry's own definition of fundamentalism in this volume is somewhat broader than the self-definitions offered by Bauder and McCune. Henry is describing a group that would include evangelicals. However, in large part, the neo-evangelicals split from the fundamentalists over social engagement.

26. Henry, *The Uneasy Conscience of Modern Fundamentalism*, 3.

27. Henry, *The Uneasy Conscience of Modern Fundamentalism*, 4.

28. Henry, *The Uneasy Conscience of Modern Fundamentalism*, 5.

attitude that led to the split between fundamentalists and conservative evangelicals. Dealing with the ills of the world relies upon cooperation between churches—often among those with varying beliefs—in order to have sufficient political capital to make changes in the world.

Such alliances have happened between fundamentalists and the secular world for purposes such as Prohibition and stopping pornography.[29] Some contemporary versions of fundamentalism, such as the streams that have participated in the so-called Moral Majority, find political engagement as a central aspect of fundamentalism.[30] However, Henry's critique represents one of the most common such criticisms against fundamentalists both as a sociological movement and as a theological group. There is within fundamentalism a tendency toward world-denying asceticism, which, while not a central tenet of the faith, is observable from the outside. Sometimes this is characterized by a seeming rejection of basic precepts of contemporary science.[31] Adherence to a young earth and the central role of the ancient text of Scripture seem to alienate fundamentalists from contemporary culture. Many observers make such statements as a form of criticism, though to some fundamentalists other-worldliness is considered a compliment.

At its best, fundamentalism is characterized by an earnest concern for doctrinal consistency with historic orthodoxy as evidence of the gospel's outworking.[32] This righteous concern for theological integrity has been a great benefit to world Christianity by maintaining a robust defense of supernaturalism, the trustworthiness of Scripture, and the purity of the visible church. To use a military metaphor, fundamentalists have been faithful in holding the line against the assaults of theological modernism in many forms. These positives have not come without costs; the most tragic costs have been a disposition toward separatism and disengagement from important, if secondary, theological issues such as environmental ethics. Disengagement from an important social cause

---

29. For example, Russell Moore references the unlikely alliance of feminists and fundamentalists in trying to root out pornography. Moore, *Onward*, 17.

30. The centrality of political engagement to fundamentalism was, for example, the view of Jerry Falwell, who was instrumental in the rise of the so-called Moral Majority. Williams, *God's Own Party*, 171–79.

31. Karen Armstrong, *The Battle for God*, ix. The reality of this is much more nuanced than critics of fundamentalism tend to allow. Most fundamentalists do not reject science *per se*, but rather scientism and certain philosophically-driven conclusions from science.

32. Bauder, "Fundamentalism," 29.

has been the chief failure of fundamentalists toward the environment, though their critics accuse them of antipathy toward environmental ethics.

CRITIQUES OF FUNDAMENTALIST ENVIRONMENTAL ETHICS

The difficulty with accurately discussing the Christian fundamentalist perspective on environmentalism is, in large part, due to the lack of primary sources. For a variety of reasons, participation by fundamentalists in formal academic pursuits tends to be limited, and, in general, fundamentalists reject environmentalism as a subject of interest. There are likely several reasons for this, which would all be difficult to prove absolutely, but a premillennial eschatology that assumes a total destruction of the created order seems to contribute significantly to the disinterest in environmental preservation.[33]

Though eschatology is a dominant force in the fundamentalist position on environmental ethics, it does not encompass the whole of the perspective. The doctrines of revelation, creation, and anthropology also come into play in fundamentalist environmental ethics as in other streams. The way they are combined helps to explain the relative silence on environmentalism. A significant danger in dealing with the fundamentalist perspective on creation care is that, due to the fundamentalists' relative silence, they tend to get excoriated or caricatured. Also, there is a pattern of evangelicals being described as fundamentalists, which this chapter is, in part, designed to correct. The conflation of conservative evangelicalism and fundamentalism is frequently found in non-evangelical sources, and progressive evangelicals like Michael Northcott sometimes commit the same error.

In an article on eschatology and environmental ethics, Northcott, with more venom than is typical for a scholar, attempts to explain the fundamentalist perspective, but without providing a clear definition of "fundamentalism."[34] At points in his essay, he associates voting for

---

33. Northcott, "The Dominion Lie," 94–95. In one case, a dispensational seminary professor wrote on a paper dealing with a biblical basis for creation care, "It's like shuffling deck chairs on the Titanic." Other personal interactions I have had with individuals who would self-identify as fundamentalists have been similar.

34. This may seem like an undue criticism of Northcott on the surface, but throughout the essay he accuses his subjects of lying, of misreading Thessalonians,

George W. Bush with Christian fundamentalism.[35] However, since many Americans of all theological persuasions also voted for Bush, this is not a suitable litmus test. As noted above, Northcott also associates fundamentalists with dispensationalism.[36] There is clearly overlap between fundamentalism and dispensationalism, but it is inappropriate to collapse the categories.[37] Northcott also aligns fundamentalists with theonomy, but this is questionable, since in his list of those that favor theonomy, he includes Francis Schaeffer alongside Rousas Rushdoony and Gary North.[38] There is a definitional vagueness in the article that allows Northcott to make the claim, but it is not well supported.[39]

Setting aside some of these concerns, the chief characteristic Northcott associates with Christian fundamentalists is a suspicion of environmentalism as a creeping neo-paganism.[40] He argues the resistance to neo-paganism is merely a guise to permit selfish enjoyment of private property by pushing back against state control.[41] Unfortunately, for Northcott, repeated calls by environmentalists for Christians to return to a form of neo-paganism do not help his case.[42] Northcott is correct that conflation of environmental ethics with nature worship is an exaggeration; however, history provides evidences of leftward drift by formerly orthodox theologians.[43] Northcott undermines his own critique because

---

and of supporting corporatism instead of Scripture. Northcott, "The Dominion Lie," 91–94, 100. In fact, this article is uncharacteristic of Northcott's tone in other works, which is what makes it worthy of note.

35. Northcott, "The Dominion Lie," 90.

36. Northcott, "The Dominion Lie," 94–95.

37. Northcott himself recognizes this as he finds both pre- and post-millennialists within the fundamentalist, anti-environmentalist camp. Within the essay, which is not Northcott's best work, there seems to be some internal inconsistency. Northcott, "The Dominion Lie," 104.

38. Northcott, "The Dominion Lie," 97.

39. Some critics of conservative Christians appear to argue that believers who see God's moral law as normative for all and the best path to holistic human flourishing are also theonomists. It appears this is how Northcott uses the term.

40. Northcott, "The Dominion Lie," 103–04.

41. Northcott, "The Dominion Lie," 104.

42. For example, Zaleha and Szasz, "Why Conservative Christians Don't Believe in Climate Change," 23–24.

43. The liberalizing trend of many environmental ethics is documented in Stoll, *Inherit the Holy Mountain*.

he attacks fundamentalists on the wrong fronts and fails to define the category adequately.

Despite Northcott's failed critique, however, there is room for critique of the fundamentalist position on the environment. There are at least four points of potential overemphasis within fundamentalist theology regarding environmental ethics. These do not preclude a sufficient environmental ethic; however, they do point to some problematic themes that are reasonably common within fundamentalism. First, some fundamentalists have an exaggerated expectation of the application of God's sovereignty in the created order. Since the majority of Christian fundamentalists hold to some form of young-earth creationism, it naturally follows that they have a greater sense of God's immediate involvement in creation history than those holding some version of an old earth. God's immanence in history, in some interpretations, supports his sustaining work as likely to overcome human degradation. This can sometimes limit concern for perceived ecological crises among fundamentalists because they believe a gracious God will intervene before humans self-destruct. Second, there is sometimes an overemphasis on imminent eschatological anticipation. Whatever their stance on the millennium, fundamentalists tend to hold out hope of God's near-term intervention in the world to bring an end to this age and usher in the age to come.[44] In some cases, this can limit concern for the long-term health of the environment. Third, there is often, though not always, a form of dualism within fundamentalism that values spirit over physical things.[45] Thus saving souls is a much higher priority than stopping global warming or any other environmental crisis.[46] Critics outside fundamentalism often over-accentuate this aspect of fundamentalism, but it bears consideration nonetheless. Fourth, there is sometimes an overemphasis on grace, which can lead to a sense of cheap grace. Fundamentalists believe it is safe to focus on the greater concerns of doing missions and saving souls to the exclusion of creation care because God will remake the entire earth in a new act of creation,

---

44. Fowler, *The Greening of Protestant Thought*, 49–57.

45. Cooper, "Dualism and the Biblical View of Human Beings (1)," 13–16. Cooper's article is a critique of dualism in Christianity. As with other theological leanings among fundamentalists, the particular error of dualism is difficult to directly discover. However, from experience as a young fundamentalist, I was taught to value the soul to the exclusion of valuing the body. Note that Wirzba picks up on this concept and debates it at several points in his recent volume, *From Nature to Creation*, 20–22.

46. Wilkinson provides evidence of this view from interviews conducted in her research. Wilkinson, *Between God and Green*, 98–99.

like the original creation of the world.[47] This fourth over-emphasis is related to the first, but reflects the expectation of God's counteracting of human activity by grace more than trust in his providence in sustaining the world.

These four elements are often exaggerated in critiques of fundamentalism. However, they are often present in some form: the stereotype has a basis in reality.[48] Taken in the best light, each of the emphases has strong positive elements. It is good to believe in God's imminent action in creation, Christ's soon return, the importance of spiritual things, and the value of the salvation of souls over earthly gains. Taken to an extreme, however, these tendencies can cause a neglect of important stewardship responsibilities.

Robert Fowler notes, "Both relative neglect and relative denial [of environmental issues] have flourished within Protestant fundamentalism . . . Here as in many other areas, Protestant fundamentalists do not drift with the flow and do not intend to do so."[49] The lack of fundamentalist participation in environmentalism based on their Christian confession is largely a result of prior political commitments, not a result of a deficient theology incapable of generating energy for a positive attitude toward the environment, as the next section shows.[50]

A GENERIC FUNDAMENTALIST ENVIRONMENTAL ETHICS

Fundamentalists are a variegated species, much like all of the theological types described in this project. However, to allow for the clearest and most concise presentation of fundamentalist doctrines, this chapter only discusses the dispensational stream of fundamentalism. It is by far the

47. Chafer, *Systematic Theology*, 4:401.

48. I have found some of these errors to be more consistent with fundamentalists outside of academic circles. Much of the work of fundamentalist academics attempts to correct this sort of error. However, the social egalitarianism of fundamentalism, which tends to prioritize calling to ministry over seminary degrees, allows some versions of folk doctrine to flourish in churches.

49. Fowler, *The Greening of Protestant Thought*, 22.

50. In an article analyzing the 1993 General Social Survey, two sociologists argue that the resistance to environmentalism by fundamentalists (their definition may include some evangelicals) is primarily driven by political associations and economic concerns rather than direct doctrinal markers. Eckberg and Blocker, "Christianity, Environmentalism, and the Theoretical Problem of Fundamentalism," 343–55.

most caricatured segment of the fundamentalist population among environmentalists.[51] Adherents of dispensationalism, such as Charles Ryrie, argue that only dispensationalism is consistent with the literal reading of Scripture, which is central to the discussion on revelation below.[52] Additionally, anticipation of a complete destruction of the present created order, which is representative of the dispensationalist position, is the most difficult position to reconcile with an environmental ethics. For these reasons, the following discussion will be limited to dispensationalist fundamentalists. Consistent with the definition of theological perspective for environmental ethics, this chapter will outline the response to the four doctrinal questions by fundamentalist theologians and present them under the headings of revelation, creation, anthropology, and eschatology.

Revelation

To a greater degree, perhaps, than the other theological perspectives discussed so far, the relationship between authority and revelation is central to understanding a fundamentalist theological perspective for the environment. Considered in the categories of the Wesleyan Quadrilateral, the dominant influence is special revelation through Scripture. Since the objective truth of Scripture was unerringly promulgated by God, it is absolute, and all other sources of authority must be subordinated to it. Tradition has a place inasmuch as it conforms to Scripture; human experience, including empirical science, must do the same; reason must also be conformed to Scripture, as it was negatively influenced by the human fall in Genesis 3. In treating the doctrine of revelation from a fundamentalist perspective, this section treats Scripture with a more extended treatment, dealing with the roles of tradition, reason, and experience in more summary fashion.

A common critique of the fundamentalist understanding of Scripture is that it is "literal."[53] When used by critics like Harvey Cox, the term "literal" means wooden and obtusely interpreted. However, critics overstate by making this accusation, since the supposedly literal reading of Scripture would never, with integrity, interpret passages like Ps 91:4

---

51. E.g., Williams, *God's Own Party*, 15.
52. Ryrie, *Dispensationalism Today*, 86–109.
53. E.g., Cox, *How to Read the Bible*, 205–07.

to indicate that God actually has feathers and wings. Dwight Pentecost notes, "It is recognized by all that the Bible abounds in figurative language... However, figures of speech are used as means of revealing literal truth."[54]

The use of the term "literal" in describing fundamentalist interpretation is not without merit, however, as Pentecost argues, "Perhaps the primary consideration in relation to the interpretation of prophecy is that, like all other areas of Biblical interpretation, it must be interpreted literally. Regardless of the form through which the prophetic revelation is made, through that form some literal truth is revealed. It is the problem of the interpreter to discover that truth."[55] Pentecost is, therefore, calling for so-called literal interpretation, but with the understanding that there is interpretation involved. As Pentecost argues,

> In the literal method Scripture may be compared with Scripture, which, as the inspired word of God, is authoritative and the standard by which all truth is to be tested... One does not have to depend on intellectual training or abilities, nor upon the development of mystical perception, but rather upon the understanding of what is written in its generally accepted sense. On such a basis can the average individual understand the Scriptures for himself.[56]

There is clearly more to the method of interpretation Pentecost is proposing than a bare literalistic reading. Rather than using the much abused term "literal," it might be better to describe the fundamentalist perspective on revelation as an argument that the Bible, which is the chief form of special revelation, has objective content or, to use another term, is *propositional*.[57] In contrast to other theories of revelation, many of which focus on the experience of the human author or the response of the reader, this approach assumes there is discernable content in the text itself, recorded for the benefit of later readers. It also presumes that such propositions can be transmitted into truthful statements that can, in turn, be communicated as consistent doctrines over time. This idea has epistemic connotations.

---

54. Pentecost, *Things to Come*, 12.
55. Pentecost, *Things to Come*, 60.
56. Pentecost, *Things to Come*, 12.
57. Martin, "Special Revelation as Objective," 61–73.

In 1994, Robert Fowler argued that Protestant Christian environmentalists were nearly all foundationalists.[58] That is likely less true today than it was at that time, given the significant growth of the ecotheological strains of liberation theology and process theology. However, fundamentalists especially, along with evangelicals, are firmly within the foundationalist camp. As Carl Henry argues in *God, Revelation, and Authority*, theology as it is practiced by many conservative Christians has a scientific method that relies on the truthfulness of Scripture and the sufficiency of human reason to rightly process that truth (within limits). Henry's theology also relies on the human ability to express the truth of Scripture in objectively true, verifiable doctrines.[59] According to critics, unlike liberal forms of theology that tend to adapt Christianity to contemporary social and scientific ideas, fundamentalists derive "from natural science their understanding of the nature of truth and [insist] that the truth of the Bible must be this kind of truth."[60]

The fundamentalist position is that the Bible contains propositional truth; indeed it is comprised of propositional truths without a mixture of error. This entails the belief that truth does not change over time. According to post-liberal George Lindbeck, "For a propositionalist, if a doctrine is once true, it is always true, and if it is once false, it is always false."[61] For a fundamentalist, the revision of a doctrine is necessarily an admission of past error.[62] Reluctance to alter doctrine sets them in strong opposition to theological liberalism and ecotheology, which actively seek to reconcile Christian doctrines to present conditions. In many cases, the resistance to reformulating doctrines—including the doctrine of revelation—has helped fundamentalists sustain traditional orthodoxy in the face of cultural challenges. However, at the same time, such resistance sometimes moves from contending for the faith to being contentious about the faith. It also has the potential to allow fundamentalists to overestimate their objectivity and fail to recognize their own significant cultural misconceptions.

---

58. Fowler, *The Greening of Protestant Thought*, 9.

59. Henry, *God, Revelation, and Authority*, 1:202–43. Recognize that Henry is not a fundamentalist but an evangelical, but his position on revelation is consistent with a fundamentalist understanding.

60. Barr, *Fundamentalism*, 93.

61. Lindbeck, *The Nature of Doctrine*, 2.

62. Lindbeck, *The Nature of Doctrine*, 33.

The real danger of the propositionalist approach is, as Richard Lints notes, the digression into the fundamentalist fallacy, which is "the conviction that God reveals himself outside of a cultural setting to communicate timeless truths to people who themselves are not influenced by their own cultural setting."[63] Unfortunately, even those most rigorous attempts at developing a biblical worldview cannot escape the effects of the surrounding culture.

Even as fundamentalists critique modernistic anti-supernaturalism, they often hold to a very modern approach to Scripture. George Marsden notes that the fundamentalist hermeneutic simultaneously affirmed both supernaturalism and a Baconian scientific mindset.[64] He writes, "[Fundamentalists] were absolutely convinced that all they were doing was taking the hard facts of Scripture, carefully arranging and classifying them, and thus discovering the clear patterns which Scripture revealed."[65] This approach, which also assumes a high degree of perspicuity of Scripture, tends to make fundamentalists more adamant about their readings and more likely to divide from those who differ with them on various points of doctrine, including eschatology. This "scientific" reading of Scripture, however, despite its usefulness in resisting liberalization of doctrines, is very modernistic and thus a part of the very same cultural milieu it is designed to refute.

The strength of such a modernistic approach to hermeneutics is that it recognizes the integrity of Scripture and the ability of the Bible to convey truth across time and culture. It allows for the unique, supernatural character of Scripture to be celebrated and to bring its full weight to bear on ethical issues. Hearing the very voice of God speaking about contemporary situations, without error and with clarity, is an essential element of the fundamentalist perspective on special revelation.[66] In some few cases, the doctrine of inerrancy is accompanied by adherence to the mechanical dictation theory, where the human authors of Scripture were merely scribes for the inspired language of the Holy Spirit. However, historian Nathan A. Finn shows that adherents of mechanical dictation theory were a tiny and largely ostracized minority.[67] Even within the

63. Lints, *The Fabric of Theology*, 8.

64. Marsden, *Fundamentalism and American Culture*, 55–56.

65. Marsden, *Fundamentalism and American Culture*, 56.

66. Marsden briefly touches on the rise of the emphasis on inerrancy from this scientific approach: Marsden, *Fundamentalism and American Culture*, 56–57.

67. Finn, "John R. Rice, Bob Jones Jr., and the 'Mechanical Dictation' Controversy,"

fundamentalist camp there are shades of understanding of the nature of Scripture. However, fundamentalists are united around the concept that Scripture is without error in its original manuscripts and that the truths God intended to reveal are discernable to the average reader within the pages of the Bible.[68]

The high regard of fundamentalists (and many evangelicals) toward Scripture causes some critics to accuse fundamentalists of worshiping the Bible instead of God. Ernst Conradie critiques this approach to Scripture, arguing, "Some fundamentalist Christians sometimes talk as if they believe in the Bible itself, as if the Bible itself is God. They attribute divine characteristics to the Bible. The Bible is regarded as equally trustworthy, authoritative, and inspired compared to Godself."[69] His criticism is, in part, true. There is a close identification of the Bible as the word of God in fundamentalist theology, and of the Word of God as Jesus, the second person of the Trinity. This does not, however, indicate that the Bible itself is equated with God, yet fundamentalists do categorically believe that the Bible authoritatively and accurately communicates information from God. In that sense it is "equally trustworthy, authoritative, and inspired compared to Godself" because the Triune God intended it to be that way, and Scripture informs that understanding of itself. The link between the Bible and God is found in the fundamentalist understanding of inspiration.

In his *Systematic Theology*, L. S. Chafer finds evidence of the inspiration of Scripture in 2 Timothy 3:16 and 2 Peter 1:21; these two passages lead him to conclude, "The Scriptures in their entirety are effective since they are from God, God-breathed, God-given, and God determined."[70] While this does not quite equate Scripture to "Godself," as Conradie accuses, it does assign a direct role in the authorship to God, so that it may be accepted that "the New Testament writers were no less the voice of God" than the Old Testament prophets.[71] The prophets were men speaking the very words of God to their contemporary audiences; the human authors of the Old and New Testament books fulfilled a very similar function.

60–75.

68. Marsden, *Fundamentalism and American Culture*, 60–62.
69. Conradie, *Angling for Interpretation*, 73.
70. Chafer, *Systematic Theology*, 1:80.
71. Chafer, *Systematic Theology*, 1:85.

In the fundamentalist approach to Scripture, the inspiration of the Bible is not merely a sense of artistic passion generated by the work of the Spirit. Instead, it implies dual authorship; this concept is central to the fundamentalist understanding of the Christian religion. As Elmer Towns argues,

> Christianity is as credible as its foundation which is the immutable rock of holy Scripture. But this Book is no natural book, for it claims to be a divine revelation, penned by human authors. As such, the Bible has dual authorship, written by men in their own language, yet the process was supernaturally guided by the Holy Spirit so that both words and messages are without error or mistake.[72]

Conradie's critique is, therefore, partially justified. The fundamentalist does indeed believe Scripture to be "equally trustworthy, authoritative, and inspired compared to Godself" because God was directly involved in the authorial process. However, Conradie's criticism is only biting when the reader is sympathetic to claims that there are errors in Scripture and that the foundation of Christian belief should be assumed to be something other than the Bible. In short, unless one rejects Scripture as an unadulterated source of moral truth, there is no reason to be concerned with Conradie's criticism. Ignoring the disgendering of God, a fundamentalist is more likely to shout amen than cringe when faced with trusting Scripture as an authoritative word from God.

From a fundamentalist theological perspective, the notion of inspiration and all that it entails is necessary for any meaningful progress to be made in understanding the Christian religion. For example, in L. S. Chafer's *Systematic Theology*, he writes, "The inspiration of the Scriptures are assumed... As a chemist will make no advance in his science if he doubts or rejects the essential character of the elements which he compounds, so a theologian must fail who does not accept the trustworthiness of the word of God."[73] Here again the Baconian modernistic understanding resonates with fundamentalism, as Chafer likens the theologian to the scientist. The greater point, however, is that fundamentalists believe that trusting Scripture is the gateway to any proper form of Christianity.

Since Scripture is absolutely normative for Christian theology, it is therefore also the supreme norm for Christian ethics. James Gustafson

---

72. Towns, *Theology for Today*, 28.
73. Chafer, *Systematic Theology*, 1:7.

describes "the most stringent use of scripture as revealed morality," which he defines thusly: "Those actions of persons and groups which violate the moral law revealed in scripture are to be judged morally wrong."[74] Elsewhere, Gustafson specifically attributes this approach to ethics to the fundamentalists' understanding of propositional revelation.[75] This approach, however, tends to limit the fields of concerns for the contemporary ethicists, since the authors of Scripture did not have the contemporary challenges of hydraulic fracturing and nuclear waste in mind. This may help explain limited interest among some fundamentalists in environmental ethics. Whatever the reason for their silence, the supreme moral authority of Scripture is clear in fundamentalist theology. The other legs of the Wesleyan Quadrilateral—reason, experience, and tradition—are subordinate to the content of Scripture.

Despite the supremacy of Scripture, the components of experience and reason, specifically as seen in contemporary, empirical science, have a significant part to play within the fundamentalist theological perspective for environmental ethics. Unlike other theological schemas, however, fundamentalism tends to initially resist new scientific ideas that appear to conflict with existing scriptural interpretations. During a time of scientific fervor in and after the Enlightenment, Christians were slow to latch onto seemingly novel interpretations of the world. Sometimes this resistance was justifiable in retrospect. This conservative attitude led to the development of the conflict model of relationship between science and religion. In similar fashion, fundamentalism was largely formulated as a response to alleged theological capitulations of liberal theologians to science.[76] As McGrath notes, in the post-Enlightenment world, "the Christian churches were regarded as the bastions of conservative thinking, and the most significant obstacles to the process of liberation and liberalization which many wished to unleash in Western Europe and elsewhere."[77] Science, which ever focuses on the new development, changed quickly; in contrast, the propositional understanding of Scripture calls for slow or no change. This leads to reluctance to accept scientific discoveries with apparent contradictions to Scripture.

---

74. Gustafson, "The Place of Scripture in Christian Ethics," 160.
75. Gustafson, "The Changing Use of the Bible in Christian Ethics," 135.
76. Olson, *The Journey of Modern Theology*, 236.
77. McGrath, *The Foundations of Dialogue in Science and Religion*, 23.

To many fundamentalists, science is a mutable human work, while the word of God is eternal. As Chafer writes,

> Science is ever shifting and subject to its own revisions, if not complete revolutions. It reflects with a good degree of accuracy the progress from generation to generation of human knowledge. In the field of science, no human author has been able to avoid the fate of obsolescence in later periods; yet the Divine Records have been so framed that there is no conflict with true science in this or any age of human history.[78]

Thus the fundamentalist does not necessarily reject the content or implication of contemporary science.[79] Rather, fundamentalists operate under the assumption that new ideas may replace contemporary science at any time, perhaps through a Kuhnian revolution. Such a new discovery would require recasting science-centric worldviews in light of the changing scientific understanding. It is often possible for the ideas of Scripture to be reconciled to the notions of contemporary science.[80] However, if the notions of contemporary science are subject to revision, the need to reconcile them fades. In some sense, then, fundamentalism does not reject science but waits for human understanding to catch up with special revelation.[81]

This attitude, in part, helps to explain the fundamentalist silence on the issue of the environment. Reading Scripture with a view to gleaning (or creating) environmental insights is a distinctly contemporary phenomenon. Just as the earth will pass away, so will this contemporary cultural concern for the environment. In one sense, the failure to engage the issue may be evidence of merely letting a contemporary fad pass by. This cannot be a full explanation of the relative silence of fundamentalists in this area; fundamentalists did, after all, speak widely against the Harry Potter phenomenon in its heyday. The answer may lie in the fundamentalist understanding of the epistemological foundations of science.

Science as a way of describing the world is reliant upon both subjective, temporal human experience and the vagaries of human reason.

78. Chafer, *Systematic Theology*, 1:34.

79. This contra-Mercer's argument that "all fundamentalists and many, perhaps most, evangelicals as well reject scientific and historical consensus where it is perceived to contradict a literal reading of the Bible, and are resistant to critical thinking about religion." Mercer, *Slaves to Faith*, xxiv.

80. Barr, *Fundamentalism*, 92–97.

81. Chafer, *Systematic Theology*, 2:130–35.

According to a fundamentalist account of the fall, one of the impacts of sin is to distort the ability of humans to rightly reason.[82] This is not an argument that all human reason is flawed, but that human reason is bent away from objective absolute truth. Such an argument explains the need for dual authorship of Scripture and the need for Scripture to have been given by God to provide a means to understand the way of salvation. This means that science may pursue absolute truth, but will inevitably fall short of achieving it. Scripture, as illuminated by the Holy Spirit, is the only means to an objective understanding of truth as God has intended humans to understand it.[83] Science, which is the embodiment of reason and experience, functions as a source of authority within a fundamental theological perspective, but it is a role governed by meticulous comparison to the content of Scripture.

The fourth point on the Wesleyan Quadrilateral is tradition, which fundamentalists sometimes eschew publicly. The most common target of criticisms regarding the use of tradition among fundamentalists is the Roman Catholic Church. According to Chafer, "One of the greatest errors of the Church of Rome is that of making the church, and not the Bible, the immediate and final authority in all matters of divine revelation."[84] In contrast, however, "Christ disregarded the traditions of men and for this was condemned by the religious leaders of the day."[85] Thus tradition, as a category of information that has moral force, is set aside in the fundamentalist system.

The apparent rejection of tradition as a normative force within fundamentalist theology is subject to obvious criticism. Fundamentalism, particularly American varieties, redounds with tradition. A preference for maintenance of tradition is a significant contributor to the "King James only" debate and some instances of the rejection of contemporary music in lieu of hymns.[86] These are traditional cultural forms that have an authoritative role in many fundamentalist churches, and often enshrine aesthetic values of a particular cultural period. Additionally, as has been discussed above, the modernism of some fundamentalist hermeneutics tends toward an inability to recognize bias caused by the

---

82. Ryrie, *Basic Theology*, 218.

83. See for example Chafer's comments on the inability of the unregenerate to accept his understanding of the creation: Chafer, *Systematic Theology*, 7:146.

84. Chafer, *Systematic Theology*, 1:14.

85. Chafer, *Systematic Theology*, 1:14.

86. Carson, *The King James Version Debate*, 83.

interpreter's own cultural context.⁸⁷ This criticism notwithstanding, it seems fair to say that fundamentalism accepts Scripture as the supreme norm and perhaps nearly the sole norm over all matters of life and faith. As such, the doctrine of revelation is the key to understanding all other fundamentalist doctrines, including the doctrine of creation.

Creation

Determining the source of value of the created order through the doctrine of creation proves difficult using fundamentalist sources on the doctrine of creation. Fundamentalist theologians tend not to address the classical doctrine of creation separately; they focus on the question of the origins of humans specifically and only more generally on the nature of creation. This opens fundamentalists to accusations of anthropocentrism. Despite the need to look to other doctrinal heads, particularly anthropology and eschatology, for a more complete understanding of the value of nature in fundamentalist theology, the doctrine of creation helps to inform a fundamentalist theological perspective for environmental ethics.

Ryrie's *Basic Theology* includes no separate section for creation, but treats the doctrine of creation within the section titled "Man: The Image of God." Most of the chapter focuses on debunking theories of origins other than young-earth creationism.⁸⁸ There is little, if anything, in that chapter that treats the impact of human sin on the environment, God's ongoing work in creation, or other theological topics which most other systematic theologians address.

Similarly, Towns does not have a separate chapter in *Theology for Today* for the doctrine of creation. He addresses various concerns with the age of the earth but not the value of the created order. Like Ryrie, Towns considers the doctrine of creation mainly under the heading of anthropology.⁸⁹ The most in-depth discussion of God's work of creation comes under the heading of pneumatology. Towns concludes the section,

---

87. The difficulty in diagnosing personal bias due to a modernist understanding of objective truth is the foundation for rejection of the reality of objective truth by Phillip Kenneson. His critique is warranted, but his later argument indicates a rejection of the possibility of completely attaining objective truth, not its actual existence. Kenneson, "There's No Such Thing as Objective Truth, and It's a Good Thing, Too," 155–70.

88. Ryrie, *Basic Theology*, 180–88.

89. Towns, *Theology for Today*, 555–622.

"The Holy Spirit had an active role in the creative work of God. The result is the order, life, beauty, and renewal properties of the created world that would not exist in the same degree without his activity."[90] This discussion is helpful, but it lacks some of the key components necessary to promote a thoroughgoing understanding of the value of God's creation, which is a necessary element for environmental ethics.

Chafer's expansive *Systematic Theology* offers little direct data on the value of the created order. Like Ryrie and Towns, Chafer seems preoccupied with the controversy over the age of the earth and its origin. His summary of the doctrine of creation focuses on accepting a version of young-earth creationism. This is posed as a sharp dichotomy between "accepting revelation" or "disregarding revelation."[91] Belief in evolutionary theories is consigned to "godless scientists," the unsaved, and others who lack sufficient faith.[92] Chafer also treats God's work in creation briefly in his pneumatology, though this quickly devolves into a rejection of evolutionary theories.[93] There is a brief reference to Christ as Creator of all things under the heading of Christology,[94] and also a quick discussion concerning the creative work of the Trinity.[95] The fatherhood of God over creation gets a full paragraph of attention, though in these lines Chafer zooms in on the relationship between created humans and God the Creator.[96] In similar fashion to Ryrie and Towns, Chafer embeds his longest discussion of the doctrine of creation under the heading "The Origin of Man."[97] Detailed attention is given in this chapter to debunking evolutionary theory and examining Bishop Ussher's chronology for human history, which Chafer holds at a respectful distance. Little attention is directed toward God's ownership of creation, the value of creation, or the breadth of theological data in Scripture about creation apart from what may be indicated about the age of the earth.

Most theologically fundamentalist Christians hold to creation in six twenty-four-hour days.[98] These three systematic theologies provide

90. Towns, *Theology for Today*, 283.
91. Chafer, *Systematic Theology*, 7:99–101.
92. Chafer, *Systematic Theology*, 7: 100.
93. Chafer, *Systematic Theology*, 6:27–28.
94. Chafer, *Systematic Theology*, 1:342–43.
95. Chafer, *Systematic Theology*, 1:305–06.
96. Chafer, *Systematic Theology*, 1: 312–13.
97. Chafer, *Systematic Theology*, 2:130–43.
98. A 2006 article in *Science* sees a direct correlation between the fundamentalism

evidence of that bias, and help explain why the bias toward creationism continues. Creationism—the belief in a week-long version of the creation narrative at a relatively recent time—is associated with the central doctrine of inerrancy for many fundamentalists.[99] This understanding puts many fundamentalists at odds with the broader culture's understanding of cosmology, and some critics argue it fuels anti-environmentalist sentiments.[100] It is more likely that creationism distracts from arguments about the environment, but does little to fuel an anti-environmental attitude among fundamentalists.

More significant in the apathy of fundamentalists toward environmental activism is their understanding of the value of creation. This survey of the doctrine of creation in fundamentalist theology points toward a generally instrumental understanding of the value of the created order, though that is difficult to prove absolutely given the dearth of direct focus on the question. Since fundamentalists located the doctrine of creation primarily under anthropology, there is some logic in inferring that for them God's work in creation is primarily about humanity. Thus a

---

in the United States and belief in creationism: Miller, Scott, and Okamoto, "Public Acceptance of Evolution," 765–66. A recent Pew Research study focuses on religiosity in general, but uses shifting acceptance of evolution as one indicator of a decline in religion: Cooperman, Smith, and Cornibert, *US Public Becoming Less Religious*, 115–16. For some history on the connection between fundamentalism and creationism, including contemporary trends, see Bleckmann, "Evolution and Creationism in Science: 1880–2000," 151–58. Here it should be clearly noted that the fundamentalism being spoken of is the contemporary brand, which is to the right of evangelicalism theologically, though generally consistent with it. This is abundantly clear because famously many of the original fundamentalists were open to much of what the science of their day said about the age of the earth and development of species. For example, B. B. Warfield advocated for the acceptance of evolutionary theory while maintaining a staunch inerrancy. Zaspel, "B. B. Warfield on Creation and Evolution," 198–211.

99. According to an article published by Answers in Genesis, a self-proclaimed fundamentalist Christian organization, belief in a six-day creation that occurred several thousand years ago is a defining characteristic of fundamentalism. In this article, Tim Chaffey argues young-earth creationism is a necessary corollary to the doctrine of biblical inerrancy: Chaffey, "The Rise and Fall of Inerrancy in the American Fundamentalist Movement." There is a militant aspect of the argument for a six-thousand-year age of earth, exclusive of any variations, which prompts a fundamentalist professor from Detroit Baptist Theological Seminary to write a plea for realism from his more argumentative peers: Snoeberger, "Why a Commitment to Inerrancy Does Not Demand a Strictly 6000-Year-Old Earth," 3–17. For a discussion on historical interpretations of the days of creation, see Lewis, "The Days of Creation: An Historical Survey of Interpretation," 433–55.

100. Berry, "Disputing Evolution Encourages Environmental Neglect," 128–29.

consideration of a fundamentalist understanding of anthropology should help inform a theological perspective for environmental ethics.

Anthropology

The question of the value of creation is answered in large part through the clearly stated fundamentalist position on the relationship between human and non-human creation. In the fundamentalist understanding, humans are part of creation, but a special part of creation. According to Elmer Towns, humans are "God's crowning creation."[101] In his discussion of the theological significance of humanity, Towns notes that humans are part of creation, but have a unique place in creation. He also emphasizes the finiteness of humanity and the importance of understanding that finiteness for a spiritually healthy life.[102] Despite the finiteness, Towns argues, "Man's value is great, for he is, with the exception of the angels, the highest of the creatures."[103] Humanity has great value, but that value is rightly understood to be subordinate to God. According to Towns, "dominion generally appears to be a result of the image of God placed within man, but not necessarily part of the image."[104] This definition places humans in the position of having dominion over creation as a functional, but not essential, aspect of humanity. Though Towns does not develop the relationship of humans to the created order in detail, it follows that since his concept of dominion is related to the image of God, the human role in creation is similar, though subordinate, to God's lordship over creation. This permits an understanding of human ownership, and justification of instrumental use of creation, but it is far from permitting the wanton abuse of creation of which fundamentalists are sometimes accused.

Like Towns, Ryrie does little to deal with the relationship between humans and creation. In *Basic Theology*, his doctrine of creation is largely comprised of a single chapter in the section on the doctrine of humanity in defense of young-earth creationism.[105] The clearest reference to the relationship between humanity and the rest of creation comes in his treatment of the fall of man. While considering the responsibilities of the

---

101. Towns, *Theology for Today*, 555.
102. Towns, *Theology for Today*, 564–65.
103. Towns, *Theology for Today*, 565.
104. Towns, *Theology for Today*, 574.
105. Ryrie, *Basic Theology*, 180–88.

primal couple prior to the fall, Ryrie points toward the so-called cultural mandate of Genesis 1:26–28. However, he is quick to note that the cultural mandate was not transmitted to the descendants of Noah after the flood (Gen 9:1). Ryrie uses this to imply that contrary to the interpretation of theonomists and Reformed theologians, the human responsibility to exercise dominion ended when Adam ate the fruit in the Garden of Eden.[106] This seems to indicate a tendency to withdraw from culture, but it leaves open the individual's understanding of the relationship between humans and the created order.

Chafer does little to address the human-creation relationship in a post-fall condition. In discussing Adam's role toward creation, he quotes Matthew Henry at length discussing Genesis 2:15. Henry (and Chafer by implication) argues that humans had responsibility within the created order before the fall.[107] In his treatment of the effects of the fall, Chafer focuses on the immediate impact to humans and not on any change in responsibilities toward the created order.[108] At the same time, early in his chapter on the innocent state of humans, Chafer argues, "a poor environment tends to encourage all manner of evil."[109] It would be anachronistic to read the idea of clean air and water into Chafer's use of "environment" here without careful consideration; however, the sentence is written in the context of several pages expounding the aesthetic beauty of the Garden of Eden, so it seems that reading is not too great a stretch. Still, there is little instruction in Chafer's anthropology on the human-creation relationship, which speaks of its unimportance, if not Chafer's position on the issue. For a clearer view of the fundamentalist perspective on human responsibility for the environment, it seems necessary to move outside the canon of fundamentalist systematic theologies to an evangelical voice that might be trusted.

Although Geisler is not himself a fundamentalist by the definition offered in this chapter, his systematic theology resonates with fundamentalist theology.[110] Consistent with his Thomistic roots, Geisler argues for

---

106. Ryrie, *Basic Theology*, 202–03.
107. Chafer, *Systematic Theology*, 2:202.
108. Chafer, *Systematic Theology*, 2:215–23.
109. Chafer, *Systematic Theology*, 2:200.

110. Geisler regularly associates with non-fundamentalists in a way inconsistent with fundamentalism. For example, he participates in the Evangelical Theological Society. Based on a search of college level basic theology syllabi from self-identified fundamentalist institutions, Geisler's textbook is popular. Other popular books are

the uniqueness of humanity within creation.[111] According to Geisler, "humankind is the pinnacle of God's earthly creation" and "higher than the animals."[112] He assumes the concept of human dominion to be drawn directly from Scripture. For Geisler, dominion is not tyrannical rule. Rather, he writes, "It is important to note that what was granted was dominion, not destruction. God owns the world (Ps 124:8), and humans are to care for it for him (Gen 2)."[113] Thus, humans are to rule over creation and use it, but are not granted license to abuse it.

The data to develop a fundamentalist position of the role of humans within creation is limited, but it seems to point in the same general direction. First, humans are part of creation but unique within the created order. Humanity should be understood to be the pinnacle of creation. Second, based on their position within creation and their status as image-bearers of God, humans have been given permission, if not responsibility, to rule over creation. This includes using the created order for human benefit. Third, despite this authority to instrumentally use creation for human benefit, the right to despoil creation is not granted. Subsequently, the accusation that fundamentalist theology results in an attitude of destructive dominion is overstated. In placing humanity under God's rule and ensuring a biblical understanding of God's ownership of creation, fundamentalism tends to use the term "dominion," but indicates something closer to the contemporary understanding of stewardship.

Based on the doctrines of creation and anthropology, there is room in a fundamentalist theology for a positive environmental ethics. However, there is also space left for fundamentalist believers to minimize the significance of the environment based on the dearth of theological treatment of the human relationship with the non-human environment and the lack of a developed system of value for the created order. In many respects, fundamentalist theology has developed around controversies central to their identity. This explains the emphasis on the creation-evolution debate to the detriment of other issues. It also explains why answers to the question of value and human responsibility may be best answered in the oft-debated field of fundamentalist eschatology.

Ryrie's systematics and Wayne Grudem's text.
    111. Geisler, *Systematic Theology*, 635.
    112. Geisler, *Systematic Theology*, 634.
    113. Geisler, *Systematic Theology*, 634.

Eschatology

The most often-discussed element of fundamentalist theology in environmental discussions is their eschatology.[114] Many fundamentalists major on the idea of the premillennial, pretribulation rapture of the church. The pretribulation rapture has been added as an additional fundamental doctrine in some fundamentalist confessions and is used as a hiring criterion in some fundamentalist institutions of higher education.[115] According to Bauder's representation of the doctrinal definition of the gospel, it would seem that some fundamentalists may hold a particular interpretation of the end times to be an essential of Christianity. For many fundamentalists, the approved eschatology lines up with dispensational theology.

Dispensationalism is often found in separatist sects of Christianity, which are often fundamentalist in their theology. Ryrie defends dispensationalism against the charge of divisiveness, but he accepts the separatist tendencies of adherents to the theological system.[116] There is, then, a large overlap between fundamentalism and dispensationalism, though the two are not synonymous. George Marsden writes, "Yet, while there was not complete uniformity of [eschatological] belief, the intellectual predispositions associated with dispensationalism gave fundamentalism its characteristic hue."[117] Since the dispensationalist eschatology that is found in the theology of some fundamentalists is the common target of environmentalists, this project focuses on that significant stream of theology among the eschatologies of fundamentalists. In addition, because dispensational premillennialism is claimed by Ryrie to be the only eschatology that consistently applies a literal hermeneutic, which is the hermeneutic discussed above under the doctrine of revelation, this position reflects an

---

114. For example: Berry, "Disputing Evolution Encourages Environmental Neglect," 127; Curry-Roper, "Contemporary Christian Eschatologies and Their Relation to Environmental Stewardship," 157–69.

115. McCune, "The Self-Identity of Fundamentalism," 21. Also, Timothy Weber writes, "In the 1920s simple belief in the Second Coming of Christ qualified as a fundamental, but in the 1930s one might have to believe in Christ's pretribulational and premillennial Second Coming to be considered a fundamentalist," Weber, "Fundamentalism." As recently as the summer of 2015, Cedarville University and Summit University required faculty to affirm a pretribulational and premillennial rapture as part of their hiring credentials. Notably, the institutions were willing to consider non-terminally degreed applicants, which is somewhat surprising given the glut of terminally qualified candidates for positions at evangelical Christian institutions.

116. Ryrie, *Dispensationalism Today*, 78.

117. Marsden, *Fundamentalism and American Culture*, 62.

extreme point on the spectrum, and therefore a useful point heuristically for the examination of theological perspectives.[118]

The significance of dispensational eschatology should not be underestimated. In some circles, eschatology seems to dwarf other significant doctrines and dominate the worldview considerations of many fundamentalists. This is illustrated by the popularity of prophecy conferences in the beginning of the twentieth century, which still occur on a more limited basis.[119] The shift toward dispensationalism came through the prophetic conferences for several reasons: "(1) an emphasis on literal interpretation of Scripture, (2) the imminency of the coming of Christ, (3) an emphasis on evangelism and missions, and (4) a firm stand against postmillennialism with its teaching of world conversion."[120] The emphasis on the literal interpretation of Scripture and the stand against postmillennialism are significant in understanding the ambivalence of dispensationalism toward environmental ethics.

Much like the understanding of revelation expressed above, Ryrie argues for strict literalism in reading both Old and New Testament prophecies. He writes, "The prophecies in the Old Testament concerning the first coming of Christ—his birth, his rearing, his ministry, his death, his resurrection—were all fulfilled literally. There is no non-literal fulfillment of these prophecies in the New Testament. This argues strongly for the literal method."[121] The dispensationalist, then, is consistent in applying the literal approach to Scripture while others fall short of that hermeneutical standard. Differences in interpretation become, for some dispensational fundamentalists, not merely varied acceptable understandings but clearly identifiable errors. In his summary of his view on dispensations, Chafer argues, "To deny these varied divisions . . . is to cease to be influenced duly by the precise Scripture which God has spoken."[122] So, for many fundamentalist dispensationalists, there is no other eschatological option

---

118. Ryrie, *Dispensationalism Today*, 89.

119. For a brief illustration of the connection of Christian fundamentalism, dispensational theology, and the Niagara conferences, see Harris, *Fundamentalism and Evangelicals*, 23–26. Also, Marsden's description is helpful in this regard: Marsden, *Fundamentalism and American Culture*, 51–62. Ryrie notes, "Dispensationalism . . . originally entered the stream of American fundamentalism through the prophetic conferences," Ryrie, *Dispensationalism Today*, 82.

120. Ryrie, *Dispensationalism Today*, 82.

121. Ryrie, *Dispensationalism Today*, 88.

122. Chafer, *Systematic Theology*, 7:123.

that is consistent with Scripture. This adamant position has strengthened the idea in popular culture that a premillennial eschatology is an essential aspect of conservative Christianity.

For many critics of conservative Christianity, their understanding of fundamentalist and evangelical eschatology is based more on popular culture than any academic research. This misperception of the eschatology of conservative Christians has been influenced largely by the popularity of the Left Behind series by Tim LeHaye and Jerry Jenkins, with its accompanying pop culture franchise.[123] Many Americans understand the Left Behind series to be a biblically faithful, if dramatized, interpretation of prophetic events.[124] Many non-fundamentalist Christians, however, reject this eschatological vision and the other-worldly hope it promotes. However, the large volume of sales of diverse products in the Left Behind franchise indicates that a significant demographic, likely of fundamentalist and evangelical Christians, find the eschatology presented to be acceptable entertainment.

Although at times debates over the merits of dispensationalism focus on eschatology, the system encompasses much more than a single doctrine, representing a holistic means of reading Scripture and understanding life. Typically, there are seven dispensations recognized in dispensationalist thinking. In the first chapter of his *Systematic Theology* in his discussion of bibliology, Chafer defines a dispensation, writing, "As a time measurement, a dispensation is a period which is identified by its relation to some particular purpose of God—a purpose to be accomplished within that period."[125] He goes on to outline seven basic ages in Scripture, which encompass all of the history of the world. His account begins with the "Dispensation of Innocence," which lasted from creation to Adam's fall.[126] The divisions are outlined neatly, if not with a precise dating, ending with the "Dispensation of Kingdom Rule." It is this final dispensation

---

123. Kilner reviews the media franchise, which has grown since 2012, outlining the underlying theological leanings of the authors. He also comments on the popular phenomenon, which some critics (particularly those from mainline Protestant denominations) assume are representative of evangelical and fundamentalist Christian theology. Kilner, "Left Behind," 52–58.

124. Frykholm, *Rapture Culture*, 13–37.

125. Chafer, *Systematic Theology*, 1:40.

126. Chafer, *Systematic Theology*, 1:40.

that "ends with the creation of a new heaven and earth" that represents the *telos* of all of history.[127]

In one sense, dispensationalism is more than eschatology, but in another sense, the constructed system reduces all of theology to eschatology. By emphasizing the temporary nature of this age and the apparent deterioration of conditions supporting the coming apocalypse, dispensationalism has the potential to minimize the significance of the present age in light of the divine action that is to come at the end.[128] In other words, it can lead to a relative inaction in pursuing temporal justice. It can lead to reading much of the Old Testament as an account of a theologically distant age when God was working in a different way than now. The history documented in Scripture is seen as largely a buildup to the grand events that will come at the end of time; thus everything becomes eschatological.

Dispensational eschatology typically sees a complete destruction of the present created order and an entirely new creation event where God creates the new heavens and the new earth. As Norman Wirzba demonstrates through his failure to cite sources when describing the position, it is remarkably hard to find the position documented.[129] It is more common on "this car will be unmanned in event of rapture" bumper stickers than in print.[130] It was, however, most popularly read in Hal Lindsey's creative interpretation of prophesy, *The Late Great Planet Earth*.[131] Even Lindsey, however, is unclear on what the destruction and new creation will look like, since the biblical data is sparse.

Dwight Pentecost defends the dispensational perspective in his volume, *Things to Come*. He deals with the destruction of the earth under the heading, "The Purging of Creation," referring to the destruction of the earth, using the word "dissolution," which is found in 2 Peter 3:10 while

---

127. Chafer, *Systematic Theology*, 1:40–41.

128. In his introduction to the large section on eschatology, Chafer writes that eschatology "is concerned with things to come and should not be limited to things which are future at some particular time in human history, but should contemplate all that was future in character at the time its revelation was given." Chafer, *Systematic Theology*, 4:255.

129. Wirzba, *From Nature to Creation*, 21.

130. Zaleha and Szasz, "Why Conservative Christians Don't Believe in Climate Change," 25.

131. Lindsey and Carlson, *The Late Great Planet Earth*, 178–79. See also, *Bible Truths for Christian Schools*, 97.

discussing purging instead of absolute destruction.[132] However, Pentecost makes it clear in his discussion of the creation of the new heaven and new earth that he is not merely referring to purging, but total destruction. He writes, "After the dissolution of the present heaven and earth at the end of the millennium, God will create a new heaven and a new earth (Isa 65:18; 66:22; 2 Pet 3:13; Rev 21:1). By a definite act of creation God calls into being a new heaven and a new earth."[133] Thus Pentecost presents a complete destruction of the present creation, which is euphemistically called a "purging," with a subsequent re-creation presented in the same terms as the Genesis 1 account of creation.

It is significant to note that Pentecost relies heavily on 2 Peter 3:10–13 for support of the idea of a complete destruction of the present creation. He does cite Mattew 24:35, Hebrews 1:10–12, and Revelation 20:11 for additional support, but the support for a fiery fate of the created order is mainly read from Peter's epistle.[134] This is because 2 Peter 3 is the clearest passage on the method for the destruction of the heavens and earth.

Similarly, in his accounting of eschatology, Chafer, who preceded Pentecost, assumes that roughly the same set of passages Pentecost references clearly describe absolute destruction of the present earth. The plain meaning of these texts is deemed to be so obvious that Chafer offers the references without commentary.[135] He does, however, provide some explanation of the coming new creation event, arguing, "Of all the final works of God, none could surpass the creation of a new heaven and a new earth . . . Though only the angels may have witnessed the creation of the present order, all living creatures will observe the final act of creation."[136] Since the world will be catastrophically destroyed, a new creation event must occur that is analogous to the one described in Genesis 1.

However, there is additional evidence to be considered regarding the prediction of a destruction of the present earth in 2 Peter 3.[137] The discussion of this passage is complicated by the presence of a textual issue, which may obscure the meaning of Peter's comments on the new

132. Pentecost, *Things to Come*, 551–53.
133. Pentecost, *Things to Come*, 563.
134. Pentecost, *Things to Come*, 552.
135. Chafer, *Systematic Theology*, 4:400.
136. Chafer, *Systematic Theology*, 4:401.
137. This textual issue has been discussed in great detail and some humor in Emerson, "Does God Own a Death Star?," 281–94.

earth. The KJV renders 2 Peter 3:10 as saying that "the heavens shall pass away with a great noise, and the elements shall melt with fervent heat, the earth also and the works that are therein shall be burned up." In contrast, the ESV translates the same phrases as, "the heavens will pass away with a roar, and the heavenly bodies will be burned up and dissolved, and the earth and the works that are done on it will be exposed." A major difference exists between the rendering of the same Greek word, στοιχια, as either "elements" (KJV) or "heavenly bodies" (ESV). The first case implies basic material structure, while the second implies objects within the larger whole. This disparity sets up a difference in the nature of destruction between worldviews. Another clear difference exists between the two translations, which invites the question how being "burned up" can be confused with being "exposed," since the Greek words for the two are significantly different.

The key to the debate is in recognizing the textual variance that explains the difference between the creation being "exposed" and its being "destroyed." Some fundamentalists are skeptical of the textual criticism that undermines the idea of destruction in the passage. The *Textus Receptus*, upon which the KJV relies, uses the word meaning "will be burned up." The *Codex Vaticanus* and *Codex Sinaiticus* both contain the word which could be rendered "will be found."[138] According to Wolters, every major critical edition of the New Testament text since the late nineteenth century has used the latter word because it exists in the earliest manuscripts.[139] On the other hand, Metzger contends that although many of the oldest manuscripts have ἑυρεθεσεται, there is a high degree of doubt about its veracity—this despite the fact that the UBS committee chose the ἑυρεθεσεται reading. According to Metzger, the large number of textual variants and relative lack of consistency leave the wording questionable, and the majority of the variants tend to imply dissolution of the earth.[140]

138. Bauckham, *Jude, 2 Peter*, 303; Danker, "2 Peter 3:10 and Psalm of Solomon 17:10," 82–86; Schreiner, *1, 2 Peter, Jude*, 357–87; Wenham, "Being 'Found' on the Last Day," 477–79. See also Clark, *New Heavens, New Earth*, 232–33. Clark supports the reading of destruction. Green, *Jude and 2 Peter*, 324–35, supports the textual variant ἑυρεθεσεται but retranslates the passage to say the "heavens and earth will pass away with a roar and the elements will be destroyed by burning, even the works that are discovered in it," 330–31. This circumvents the problem. It should be noted that Green's discussion of re-creation versus renewal is somewhat confusing. Green's commentary could be taken as support for either the destructionist or restorationist perspective.

139. Wolters, "Worldview and Textual Criticism in 2 Peter 3:10," 405–13.

140. Metzger, *A Textual Commentary on the Greek New Testament*, 363–637. The

In other words, there is textual support for either word, which makes the translators' choice dependent on something other than just the text.

As Wolters argues, choosing to retain the notion of destruction in the 2 Peter 3 passage, when revealing seems more appropriate to many, relies upon and affirms a particular worldview. That worldview presumes that creation is destined for destruction. While the KJV translation was based on the best textual evidence available at that time, continuing with that translation has been the subject of a great deal of debate. Some of the concern among the so-called KJV-only proponents, nearly all of whom are fundamentalists, is that changing translations requires changing doctrines.[141] In the case of the eschatological implications of 2 Peter 3, at least, they may be correct regarding textual criticism being a basis for a change. As it stands, an eschatology that anticipates the coming destruction of the earth and subsequent re-creation has the potential to permit a solely instrumental view of the value of the environment, as critics of premillennial eschatology are ready to note.

Among the most hated men in the history of environmental ethics is James Watt, the Secretary of the Interior in Ronald Reagan's administration. Watt, a Pentecostal Christian, is most famously quoted as explaining his pro-development policies, "I do not know how many future generations we can count on before the Lord returns."[142] The assertion in most of the places that Watt is quoted is that his belief that the world will be destroyed by God at the end of this age limited his environmental ethic. In a narrow sense, this is possibly true. It is much like the difference in attitude toward the treatment of an heirloom mug versus a disposable paper cup; the presumed destruction of the created order can reduce the ethical concern for the treatment of the created order. One is much more likely to carefully carry an heirloom mug than a paper cup. This is the potentially negative side of a fundamentalist creation care. As one fundamentalist blogger writes, "All of nature and everything man has created will be completely destroyed . . . If the world is going to be 'dissolved,' there is no need for us to become too attached to it."[143] This is an attitude

---

textual apparatus of the UBS 4th edition lists nine different variants, with the largest number of witnesses supporting the two main variants εὑρεθεσεται and κατακήσεται.

141. Carson, *The King James Version Debate*, 62–66; Joe Stowell, "Foreword," viii–x.

142. For example, this is quoted in Bouma-Prediger, *For the Beauty of the Earth*, 71. However, Watt has disputed the account.

143. Todd Strandberg, "Bible Prophesy and Environmentalism." To be fair, Strandberg presents the legitimacy of good stewardship, but the short-term attitude is

common among some Christians, which has been recorded in the theology of songs and thus proclaimed vigorously in churches across America.

The destructionist attitude is codified in several Christian hymns. The hymn "This World is Not My Home," written by Albert Brumley, describes the expectation of a home apart from the present creation. The first stanza reads, "This world is not my home / I'm just a-passing through / My treasures are laid up / Somewhere beyond the blue."[144] Brumley put his escapist eschatology on display and it is now printed in hymnals to be used in worship. One can justify the basic statements of the song with Scripture, but the thrust of the song seems to be that Christians must look forward to escape from the world rather than improving the world they are in. An even more popular Brumley hymn that expresses an escapist notion is "I'll Fly Away." This lyrics of this song read, "Some glad morning when this life is o'er / I'll fly away / to a home on God's celestial shore / I'll fly away. / I'll fly away, O glory, I'll fly away./ When I die, hallelujah, by and by, I'll fly away."[145] Hope, according to this hymn, is not found in the redemption of all creation, but in human escape from the corruption of the created order. In similar fashion, the song by Alphus LeFevre, "The Old Gospel Ship," calls for a rejection of the goodness of this world and a longing for something other than this world. The lyrics, sung even by Johnny Cash, read: "I'm gonna take a trip / In the old gospel ship; / I'm going far beyond the sky. / I'm gonna shout and sing, / Until heavens ring, / when I'm bidding this world goodbye."[146] All three of these hymns, written in the early to mid-twentieth century, reflect the dispensationalist hope of escape from this world to heaven, which seems to be something different entirely from the present creation.

To be clear, none of these songs condone the abuse of creation. However, they do invoke a short-term attitude toward the created order. The soul will pass on to heaven, which is a place entirely distinct from the present earth. Therefore, ethical questions about the value of the present creation are secondary. Such an attitude is often labeled dualistic by critics, which it may be.[147] This attitude appears in some folkish versions of fundamentalism that have adopted a disembodied heavenly existence for

---

evidenced by his quote here. However, it is also fair to note that Strandberg's eschatology is the foundation of his rejection of environmentalism.

144. Brumley, "This World Is Not My Home," 485.
145. Brumley, "I'll Fly Away," 601.
146. LeFevre, "The Gospel Ship," 521.
147. Wilkinson, *Between God and Green*, 98–99.

the human soul.[148] However, in most cases, the emphasis in fundamentalist theology on the salvation of the human soul above all else simply minimizes the value of good works in exemplifying Christ's life and ministry in the world, without denying the goodness of creation. Thus, it is an issue of imbalance rather than outright error.

Some scholars argue that premillennial eschatology is a theological block to environmentalism. For example, Timothy Weber argues, "because they look at the current age as hopelessly doomed, premillennialists as a rule do not engage in long-term programs for social betterment, but rather emphasize the need to evangelize before the Second Coming."[149] Similarly, R. J. Berry asserts that the position of "premillenarianism and the destruction of this present world" is the only way a Christian can avoid the implication of Scripture that "creation care is not an optional extra for enthusiasts, but is inseparable from our calling as Christians."[150]

Despite these critiques, it is not necessary that premillennial fundamentalists abandon concerns for stewardship of the environment simply because a divine destruction and subsequent re-creation of the world is expected in the future. Returning to James Watt's comments about the imminent return of Christ is helpful. On February 5, 1981, Watt responded to a statement by Congressman James Weaver of Oregon applauding Watt's desire to continue policies of sustainable use of natural resources. Watt replied that he intended to continue those policies and then amplified his basis for his position, stating,

> That is the delicate balance the Secretary of the Interior must have, to be steward for the natural resources for this generation as well as for future generations. I do not know how many future generations we can count on before the Lord returns; whatever it is we have to manage with skill to leave the resources for future generations.[151]

---

148. Wirzba, *From Nature to Creation*, 21. Wirzba critiques the idea of a disembodied eschatological state, but he cites no sources. This perception is, I think, more a folkish adaptation of Christianity than an official position within Christianity. I have met individuals who held this position. They were fundamentalists, but they were, in general, not well theologically grounded.

149. Weber, "Eschatology."

150. Berry, "Disputing Evolution Encourages Environmental Neglect," 127.

151. Arnold, *At the Eye of the Storm*, 75.

Though Watt was premillennial in his eschatology, he retained a positive responsibility for caring for the environment.[152] Simply because there is a foreseeable destruction of something does not imply that it should not be cared for. While a premillennial eschatology may not provide direct support for a robust environmental ethics, it certainly does not eliminate the possibility of such. Many people disagree with Watt's policies and with the environmental ethics of faithful, environmentally engaged Christians. But it is one thing to have different policy preferences and another to have a theological perspective incompatible with environmental stewardship.

CONCLUSION

By constructing a fundamentalist theological perspective for environmental ethics, this chapter supports the thesis of this book that the four proposed questions provide an adequate, if basic, framework for understanding someone's approach to creation care. Based on this discussion of the fundamentalist theological perspective for environmental ethics, there is sufficient foundation for a robust concern for creation care within fundamentalist doctrine. However, an eschatology that predicts total destruction of the created order and subsequent re-creation may tend to undermine the need to treat *this* earth well if the doctrine is not well-balanced with other doctrines. Additionally, resultant overemphasis on the salvation of the human soul to the detriment of good works has the potential to hamper a proper valuation of the created order. Although it does not create a gnostic dualism, as some argue, it does tend to turn the concept of "dominion" or "stewardship" toward an instrumental valuation of the created order.

The instrumental valuation of nature is driven, in part, by the conception of the human as the pinnacle of creation. The purpose in human life is to glorify God, particularly for fundamentalists by winning more souls and becoming more spiritual. Thus, again, while not requiring a negative view of the created order, fundamentalist anthropology does

---

152. Notably, Watt was charismatic and would be rejected by most thoroughgoing fundamentalists. However, his premillennial understanding of eschatology would resonate with many fundamentalists. For a more detailed account of Watt's conversion and his beliefs, see Stoll, *Protestantism, Capitalism, and Nature in America*, 190–92.

tend to undermine their value system and place human concerns at such a higher level that environmental concerns risk being lost.[153]

Concluding his analysis of Protestant fundamentalist attitudes toward environmentalism, Robert Fowler writes,

> For [fundamentalists], the Bible is not about biocentrism; nor is it even about stewardship in a traditional sense, if that implies that the human is equal with all other parts of nature. God rules nature; God gave nature to humans for their use under the care of the divine rule; and God now calls people to prepare for the end of the world as we know it.
>
> From this perspective, God does not call on us to reinterpret the Bible as a modern environmentalist manifesto in which service of nature takes precedence over all else. In fundamentalist Protestant analysis, the crisis in creation today is not about the mistreatment of nature so much as it is about human sin in all its forms. By this view, the answer lies not in an environmental movement but in preparation for the return of Jesus Christ.[154]

Thus, according to Fowler's analysis, the trouble is not antipathy toward the environment but an otherworldly concern that leads to a neglect of duties to live rightly in the present world.

Fowler's criticism carries significant weight. As can be seen in the discussions of creation and anthropology above, some fundamentalist texts on systematic theology, including Chafer's expansive eight-volume work, have little to say regarding the value of non-human creation and the proper role of humans within the created order. There is sufficient basis for a robust environmental ethics within fundamentalist theology, but the issue gets sidelined as a secondary concern at best.

The tendency to ignore the environment leaves the contemporary fundamentalist in the same place that Carl F. H. Henry described in 1947, with a social agenda that appears to support "the widespread notion that

---

153. In my own fundamentalist upbringing, I was surrounded by men and women that cared for the environment. The rural church of my childhood was filled with avid hunters and professional farmers. Both groups considered conservation of natural resources important, though it was typically on an instrumental basis. Thus, care for the environment and Christianity do not seem to be antithetical even for fundamentalist Christians.

154. Fowler, *The Greening of Protestant Thought*, 57. Note that Fowler has a broader definition of "fundamentalist" than is being used in this chapter, but the analysis remains helpful.

indifference to world evils is essential to Fundamentalism."[155] It is not that fundamentalism has failed to recognize sin in the world, Henry argues, "But the sin against which Fundamentalism has inveighed, almost exclusively, was individual sin rather than social evil."[156] Often this is due to a rejection of the methods of advancing justice in society, as they are proposed by more liberal Christians or even non-Christians, rather than a difference in understanding of what the goal actually is.[157]

It is, perhaps, fair to observe that one of the key marks that distinguish contemporary fundamentalists from evangelicals—apart from the concern for separation—is a concern for living out doctrines in public. Thus, David Bebbington points toward *activism* as one of four consistent traits of evangelicalism.[158] The groups that remained within the fundamentalist movement and did not defect to become neo-evangelicals have, it seems, lost or largely ignored that aspect of the traditional, orthodox, evangelical identity. According to Henry, "Fundamentalism in revolting against the Social Gospel seemed also to revolt against the Christian social imperative."[159] This revolt, it seems, is a stronger explanation than theological incompatibility for the failure of fundamentalists to respond positively to environmental concerns by developing a robust environmental ethics. The problem with fundamentalism with regard to environmentalism is not bad theology, but an insufficient application of the good theology the movement possesses.

Fundamentalists are an easy mark for criticism from environmentalists because of their distinctly supernaturalist perspective on Scripture and their relative silence regarding the environment. If a theologian assumes the validity of modern (in the sense of anti-supernatural) or postmodern views on Scripture, then the fundamentalist perspective seems as odd as an Amish buggy on an interstate. However, this chapter has shown that in the four doctrinal topics that form a theological perspective for environmental ethics, there are few grounds for rejecting fundamentalist theology as a support for the environment. Instead, it appears that further development of fundamentalist theology along the lines of

---

155. Henry, *The Uneasy Conscience of Modern Fundamentalism*, 5.
156. Henry, *The Uneasy Conscience of Modern Fundamentalism*, 7.
157. Henry, *The Uneasy Conscience of Modern Fundamentalism*, 16–22.
158. Bebbington, *Evangelicalism in Modern Britain*, 10–12.
159. Henry, *The Uneasy Conscience of Modern Fundamentalism*, 22.

traditional orthodoxy, consistent with the perspective outlined here, has potential to form a solid foundation for a robust environmental ethics.

The main differences between the fundamentalists and evangelicals tend to be in the application of their respective theological perspectives for environmental ethics. Although there are some theological distinctions, the next section will show it is possible to combine theological conservatism with a positive environmental ethics.

PART FIVE

AN EVANGELICAL ENVIRONMENTALISM

CHAPTER EIGHT

AN EVANGELICAL PERSPECTIVE FOR ENVIRONMENTAL ETHICS

THE DISCUSSION OF AN evangelical theological perspective for environmental ethics is complicated somewhat by ambiguity in the definition of "evangelical." The boundaries of evangelicalism, however vague, must be established before a meaningful discussion of an evangelical environmental ethics can advance.[1] Unlike the fundamentalist perspective, there is a reasonable volume of scholarly discussion on an ethics for the environment among evangelicals. However, because they refuse to alter historically orthodox doctrines to accommodate new popular responses to climate change, some critics accuse evangelicals of being indistinguishable from fundamentalists.[2] It is this misperception, in part, that the present chapter seeks to correct.

   1. Non-evangelicals who take up the task of writing about evangelical environmental ethics often express surprise by the theological diversity within the movement. For example, Danielsen, "Fracturing over Creation Care?," 119.
   2. E.g., Horrell, Hunt, and Southgate, "Appeals to the Bible in Ecotheology and Environmental Ethics," 228–31. While Horrell and company find a wide variety of Christianity and still lump evangelicals and fundamentalists together, Gottlieb has only two categories for Christians: people who agree with him and fundamentalists. Thus, there is no difference between the separatist fundamentalist and a conservative evangelical; Gottlieb, *A Greener Faith*. In a generally excellent book on evangelical environmentalism, Katherine Wilkinson conflates all non-liberal Protestants into the

In support of presenting and analyzing an evangelical theological perspective for the environment, this chapter provides a working discussion of evangelicalism. Then, it presents a generic evangelical theological perspective for environmental ethics.

DEFINING EVANGELICALISM

One of the central problems in defining evangelicalism and outlining a consistent evangelical theological perspective for environmental ethics is that there is little agreement among evangelicals about the boundaries of evangelicalism.[3] At its very basic level, it would seem that the term *evangelical* presumes a connection to the *euangelion*—the gospel. All of those who claim to be evangelical, wherever they land on the theological spectrum, would affirm this connection. However, the gospel itself has been subjected to several definitions of varying breadths. Some define "gospel" as *only* God's action in redeeming the lost through penal substitution.[4] Others tend to have a more holistic appreciation of the work of Christ in the world, understanding the gospel as a description of God's redemptive work throughout creation.[5] Then again, others who call themselves "evangelical" have tended to overemphasize the redemptive work in the broader culture and lost sight of the redemption of the elect as the central element in the gospel.[6] However this broader debate about the exact nature of the gospel is defined, it is clear that even the connection of evangelicalism with the gospel can be significantly nuanced.

Perhaps the most commonly accepted definition of evangelicalism has come from David Bebbington. In his book *Evangelicalism in Modern Britain*, he describes four characteristics of evangelicalism: *conversionism*, *activism*, *biblicism*, and *crucicentrism*. This is often called the Bebbington Quadrilateral. Conversionism is the idea that repentance from sin and turning to Christ in faith is essential to the Christian gospel. This is a conversion to a particular, substantial faith, not merely a vague assertion that

---

category of evangelicalism. Wilkinson, *Between God and Green*, 5.

3. Robert Wuthnow argues pollsters have distorted religious categories, thus making them much more difficult to discern sociologically. See particularly his discussion in the introduction: Wuthnow, *Inventing American Religion*, 10–14. See also Kidd, *Who Is an Evangelical? The History of a Movement in Crisis*.

4. E.g., Gilbert, *What Is the Gospel?*, 110.

5. Jones and Woodbridge, *Health, Wealth, & Happiness*, 7–8.

6. Wallis, *On God's Side*, 9.

Christianity entails becoming part of a contrast community.[7] "Activism" refers to a lifestyle engaged in pursuing the gospel conversion of others to authentic faith in Christ; it is directly built upon the concept of conversionism. According to Bebbington's description, activism includes social action but not *merely* actions that seek social justice. Authentic evangelical activism necessarily includes the intent of proclaiming the gospel for the purpose of conversions.[8] "Biblicism" describes evangelicals' "devotion to the Bible, [which is] the result of their belief that all spiritual truth is to be found in its pages."[9] Bebbington notes that even in the early days of British evangelicalism there were divergences in opinions about the degree and nature of inspiration, but nonetheless there was a consistently high view among those early evangelicals.[10] Eventually, the view began to deteriorate, leading to the liberal/evangelical split around the time of the First World War.[11] The final main characteristic of evangelicalism, according to Bebbington, is a central focus on the doctrine of the cross and its necessity for human salvation. Bebbington writes, "To make any theme other than the cross the fulcrum of a theological system [is] to take a step away from evangelicalism."[12] The doctrine of the cross includes both salvation through substitutionary atonement and subsequent sanctification. These concepts were inextricably linked in Christ's work on the cross.[13]

This volume uses Bebbington's definition to describe evangelicalism. Although there are other ways to define evangelical theology, there is sufficient precision and breadth in Bebbington's definition to make it helpful. Additionally, while not universally accepted, Bebbington's Quadrilateral has found its way into several theological texts, such as the recent succinct, single-volume theology by Beth Felker Jones,[14] and is even held up as the gold standard for evangelicalism in another recent volume,

---

7. Bebbington, *Evangelicalism in Modern Britain*, 5.

8. Bebbington, *Evangelicalism in Modern Britain*, 10–12.

9. Bebbington, *Evangelicalism in Modern Britain*, 12.

10. For example, see the early Baptist pastor-theologian Andrew Fuller: Spencer, "Andrew Fuller and the Doctrine of Revelation," 207–26.

11. Bebbington, *Evangelicalism in Modern Britain*, 14.

12. Bebbington, *Evangelicalism in Modern Britain*, 15.

13. Bebbington, *Evangelicalism in Modern Britain*, 15–17.

14. Jones, *Practicing Christian Doctrine*, 5.

Introducing Evangelical Ecotheology.[15] Though not without defect as a device for classification, Bebbington's Quadrilateral is practically useful.[16]

Recent abuses of the identifying framework, however, have brought its usefulness into question. Popularly, there are those who claim status as evangelicals, but have weakened their stance in numerous areas central to the evangelical faith in an attempt to reconcile traditional evangelical doctrine with contemporary environmentalism. Recently, three academics from George Fox Evangelical Seminary have tried to unite evangelicalism to environmentalism through revisionism. Their book, *Introducing Evangelical Ecotheology*, is an attempt to involve evangelicals in dialogue about the environment, moving beyond the Lynn White debate. The problem with this text is its attempt to present a questionable theological method as consistent with authentic evangelicalism. The authors claim their approach is evangelical based on a brief citation of David Bebbington's famous evangelical quadrilateral. Their implication is that the theology they are presenting meets his qualifications of conversionism, activism, biblicism, and crucicentrism.[17] A closer examination of Bebbington's categories and ecotheology reveals a significant distance between the two. As has been discussed previously, "ecotheology" is a term for the application of liberation theology to the issue of the environment. As such, the three authors, Brunner, Butler, and Swoboda, have to redefine Bebbington's characteristics to present themselves as evangelical.

While the intent of these authors is no doubt good—to get Christians to rightly value creation—the methodology used in the book is less than helpful, as it relies on a hermeneutic more akin to one used in liberation theology than one used consistently within traditional Christian theology. This approach leads them to attempt to listen to the voice of the earth within Scripture, which contrasts with the efforts of most evangelicals to listen to the voice of God in the Bible.[18] The three authors argue,

> While we embrace the Bible for its compelling revelation of salvation and justice, we must resist applying it through a model that views interpretation as timeless and transcultural. Scripture is inspired; our interpretations are not. Thus in dialogue with other Christ-followers, we want to "glean" from the Bible and

15. Brunner, Butler, and Swoboda, *Introducing Evangelical Ecotheology*.

16. For example, Bebbington's Quadrilateral does not require an evangelical to be Trinitarian, which is problematic though largely addressed by a robust biblicism.

17. Brunner, Butler, and Swoboda, *Introducing Evangelical Ecotheology*, 5.

18. Brunner, Butler, and Swoboda, *Introducing Evangelical Ecotheology*, 23.

from our traditions' insights and practices what may have been missed or left behind by dominant structures and voices.[19]

This approach to Scripture coupled with an emphasis on orthopraxy tends to push the three authors of *Introducing Evangelical Ecotheology* away from the doctrinal roots of evangelicalism.

Such an approach is by no means new to the evangelical discussion. In 1976, Donald Dayton sought to define evangelicalism as a religious group centered on religiously-motivated action that represents the application of gospel justice to society.[20] David Wells, whose theologically-bounded definition of evangelicalism Dayton rejects, argues that the real shift in the definitional understandings of the movement is the result of a desire by evangelicals most comfortable with modernism and postmodernism to remain within the political power structure of the evangelical movement without retaining the doctrinal center.[21]

The debate over the definition of "evangelical" could spill over well beyond the pages of this chapter. Even though Bebbington's Quadrilateral is helpful in defining evangelicalism, the four categories are sufficiently broad that doctrinally heterodox theologians who claim to be evangelical do so on the basis of those four criteria.[22] As a result, deciphering a consistent theological perspective based on self-identification as evangelical would prove impossible. This project, therefore, applies the term "evangelical" to self-identified evangelicals who accept that title based on their doctrinal position rather than by line of denominational descent or social activism.

Theologically conservative evangelicals tend to define the movement doctrinally. As Richard Lints notes, the neo-evanglical movement, which is American evangelicalism, arose when "it gradually dawned on the leaders of the fundamentalist movement that they had lost the battle for the soul of the nation in the theological wars of the 1920s."[23] This led to the split between the fundamentalist movement and the neo-evangelicals. Both resultant groups were somewhat theologically impoverished,

19. Brunner, Butler, and Swoboda, *Introducing Evangelical Ecotheology*, 22.

20. Dayton and Strong, *Rediscovering an Evangelical Heritage*. Dayton was willing to use the term "evangelical" in this manner despite his denigration of the term in a 1991 essay. Dayton, "Some Doubts about the Usefulness of the Category 'Evangelical,'" 245–51.

21. David Wells, "On Being Evangelical," 391–92.

22. Gushee and Sharp, *Evangelical Ethics*, xvi.

23. Lints, *The Fabric of Theology*, 43.

because they had "set themselves to the task of protecting the doctrinal message of Scripture rather than exploring it."[24] This left the new evangelical movement "drawing their identity less from an essential theological framework than from a social and cultural strategy."[25] Such a strategy has implications for the development and application of an evangelical theological perspective for environmental ethics.

A GENERIC EVANGELICAL ENVIRONMENTAL ETHICS

The truthfulness of Lints's observations can be evidenced by the failure of evangelicals to meaningfully engage and surpass Lynn White's caustic assault on the historical theological foundations evangelicals hold to be central to their identity. The response began with Schaeffer's *Pollution and the Death of Man*, but, as Sabrina Danielsen notes, there has been precious little progress in developing a helpful theological foundation for environmental ethics because theological method has, until recently, been an issue of lesser concern for evangelicals.[26] Nowhere was this better illustrated than in 2012 at the annual meeting of the Evangelical Theological Society, where numerous papers were delivered on the theme of creation care, few of which advanced the cause of creation care in any meaningful way. One of the plenary sessions was published in the *Journal of the Evangelical Theological Society*. This speech-turned-article by Russell Moore is theologically sound, but its content is little different than Schaeffer's earlier presentation.[27] The repetition of these basic themes is not a flaw in Moore's presentation, but an indication of the stagnation of evangelical scholarship on this important issue. Moore presented a basic apology for creation care because that is what the audience of evangelical scholars needed to hear.

There are at least three reasons for the relatively limited interest among evangelicals in scholarship on creation care. First, the main theological case for a theological environmental ethics was well made by Francis Schaeffer decades ago. As Schaeffer shows, it is not hard to draw a clear line from both the Old and New Testaments to valuing creation in

---

24. Lints, *The Fabric of Theology*, 43.

25. Lints, *The Fabric of Theology*, 43.

26. Danielsen, "Fracturing over Creation Care? Shifting Environmental Beliefs among Evangelicals, 1984–2010," 201–02.

27. Moore, "Heaven and Nature Sing," 571–88.

light of the Creator and being good stewards of it. This means that evangelicals publishing on the topic are largely rehashing old material. On the other hand, many of the recent books from those to the theological left of evangelicals focus on capturing the imagination and inspiring activism in churches. They continue to reinvent the environmental message to try to find a more compelling case than the so-called cultural mandate of Genesis 1:28 and the idea of stewardship. Many of the recent books from the left are also attempting to find streams of historical theology that can be recovered or reinterpreted to redeem Christianity from its allegedly anthropocentric roots. In contrast, evangelicals typically see support for their basic interpretations of Christian environmentalism where it is addressed at all in Scripture; they tend not to try to draw out concerns for the environment from historical theologians because such concerns are anachronistic.

A second reason for the limited interest is that issues such as gender issues and religious liberty have moved to the front burner. The potentially catastrophic problems of climate destabilization seem less dire than the predicted near-term social impacts of redefining marriage and enforcing progressive sexual orthodoxies. The continued association of environmentalism with population control, centralized government, and other socially and politically liberal causes makes environmentalism a divisive issue in a time when consensus on other issues seems more significant.

Third, there is a discernable pattern of individuals who agree with liberal theologians on environmental ethics moving along a trajectory toward other liberal theological positions. Stoll's historical analysis reveals that trend to be true, and recent history supports that in some cases as well.[28] Therefore, there seems to be a correlation between delving into environmental ethics and deviation from historic orthodoxy.

Granting the smaller-than-ideal number of evangelical sources for environmental ethics, there are still sufficient voices within conservative evangelical circles to develop an outline of a generic evangelical theological perspective for environmental ethics. Like much of the contemporary evangelical theological debate about any doctrinal topic, a discussion of theological perspective for environmental ethics should rightly begin with the question of revelation and the sources of authority for theology.

---

28. Stoll, *Inherit the Holy Mountain*, 2.

Revelation

Of the four doctrinal headings that organize a theological perspective on environmental ethics, there is a strong unity among evangelicals on the question of revelation. Contra the accusations of some critics, Scripture is not the only viable source of theological authority for evangelicals, but information in Scripture has primacy over other sources. In the so-called Wesleyan Quadrilateral, which originated from the theology of an evangelical, the other three angles of the quadrilateral—tradition, reason, and experience—are valid sources, but Scripture is the norming source to which all other sources are required to conform.

Tradition is a limited source for environmental ethics because there is very little environmentally-focused writing in the history of orthodox Christianity. Wesley Granberg-Michaelson argues part of the cause of the limited tradition is an excessive emphasis on the God-human relationship and insufficient emphasis on the human-creation relationship in post-Reformation Protestant theology.[29] Granberg-Michaelson may be correct in part, though a more likely explanation for the perceived deficiency of theology relating to the human relationship with the created order seems to be the fundamental societal changes that have been enabled by the Industrial Revolution. There has been a shift from the agrarian roots of society, which tended to keep a larger portion of the population relating to nature as creation on a more regular basis. There has also been a dramatic increase in the ability for humans to negatively impact the created order due to the advancement in industrial technologies.

Even if there were a clear tradition of environmentalism within Christian tradition, it would not automatically produce a robust evangelical theological perspective for environmental ethics. As Michael Bird notes, much like liberals, evangelicals do not reliably appropriate content from their theological tradition. For liberals, the rejection of tradition is due to a theory of inevitable chronological progress. For evangelicals, the role of tradition is minimized because of the emphasis on the human origins of tradition in comparison to the divine origin of Scripture.[30] There have been appeals to historical theology as a source for environmental ethics, but these have typically come through revisionist interpretations from scholars in the liberal or ecotheological streams.[31] Though there

---

29. Granberg-Michaelson, "Introduction: Identification or Mastery?," 3.
30. Bird, *Evangelical Theology*, 64.
31. For example, Conradie, *Creation and Salvation*; Santmire, *The Travail of Nature*.

are environmentally positive elements in historical Christian theology, the traditional expressions tend to be muted because they do not deal with ecology in terms that are consistently helpful for contemporary applications. Only the role of experience plays a lesser role than tradition in the development of evangelical theology.

The role of experience is highly questioned among conservative evangelicals, with some arguing experience is inherently subjective and individualist.[32] At the same time, in the environmental movement at large, experience is regularly a primary motivation in many appeals for environmental ethics. For example, Stephen Bouma-Prediger begins his environmental ethics with an appeal to wilderness experience as a foundation for valuing creation.[33] The founder of the deep ecology movement, Arne Naess, argues for the necessity of childhood exposure to wilderness for a proper environmental ethics.[34] Such experiences, however, do not provide primary theological content and are inherently personal, which makes them difficult to incorporate into a publically framed theological perspective. To mitigate the individualistic, subjectivist nature of experience's influence, Bird argues, "Experience can only be secondary and confirmatory, not primary and absolute."[35] In the evangelical theological perspective for environmental ethics, experience with creation can illustrate the cognitive understanding of the value of the created order, but it can never be the foundational impetus for an ethical norm that runs counter to Scripture.

The next leg of the Wesleyan Quadrilateral to consider is reason, which is often connected with scientific thinking in contemporary debates. In some cases, the use of scientific reasoning leads to collisions between currently accepted interpretations of Scripture and a scientific understanding of the world, such as in the infamous Scopes Trial. However, by and large, evangelicals tend to see little conflict between their religious beliefs and science.[36] The more specific debate about the relationship of scientific reason to evangelical religious beliefs is grounded first in an overall understanding of the place of general revelation in the evangelical tradition.

---

32. Jones, *Practicing Christian Doctrine*, 24–28.
33. Bouma-Prediger, *For the Beauty of the Earth*, 19–20.
34. Naess, "Access to Free Nature," 48–50.
35. Bird, *Evangelical Theology*, 73.
36. Fink and Alper, *Religion and Science*, 12–15.

General revelation is information gained through right reasoning about data observed in the world. It is something distinct from nature itself; it is the evidence of God and God's moral order in the natural world. Evangelicals allow that God reveals himself through his creation, especially through the objective created order.[37] However, opening the door for general revelation to influence theology has its own pitfalls, especially given the potential for human bias within the observation of nature to distort the interpretation of general revelation. Such an error permits sometimes significantly flawed scientific interpretations to take an inordinately large role in determining theological principles, which then drive ethical outcomes.[38] Instead, Scripture must inform the observation of nature. As Wolters argues, "The revelation of God's will in Scripture is like a verbal explanation that an architect gives to an incompetent builder who has forgotten how to read the blueprint."[39] General revelation alone, including scientific theory derived from observations of nature, is an insufficient blueprint for an evangelical theological perspective for environmental ethics.

The necessity of special revelation is significant because it provides an additional reason for rejecting a pantheistic or panentheistic doctrine of revelation, "according to which God, by virtue of his identification with nature, *necessarily* becomes manifest in nature."[40] God makes himself known through nature, but God is not salvifically discovered through nature. There is a subtle but vital difference. It is this difference which provides security from the encroachment of a pantheistic environmental ethics which so alarmed Francis Schaeffer and which presents a real danger in some other approaches to environmental ethics. Such inappropriate appropriations of authority through general revelation are avoided by evangelicals precisely because Scripture is perceived to be the supreme authority for all of life.

As some ecotheologians lament, there is insufficient basis within the canon of Scripture for an environmental ethics that exactly matches popular, contemporary environmentalism.[41] This has led some Christians to go beyond Scripture—and sometimes contrary to it—to find an acceptable

---

37. Berkouwer, *General Revelation*, 296.
38. Berkouwer, *General Revelation*, 293.
39. Wolters, *Creation Regained*, 39.
40. Berkouwer, *General Revelation*, 316.
41. Habel, "Introduction," 25–36.

environmental ethics. Evangelicals, due to their reverence for the authority of the Bible, have tended to be more restrained in their environmentalism. Thus, when environmentalists call for abortion as a potential solution to overpopulation, evangelicals have balked.[42] Evangelicals are both constrained to act for the benefit of creation and restrained from making that the sole or even primary focus of their holistic ministries because of the authority of Scripture. A reliance on Scripture is central to an evangelical theological perspective for environmental ethics.

In fact, within the American context, the acceptance or rejection of the accuracy and authority of Scripture largely differentiates so-called progressive evangelicals like Dayton and Strong from the evangelicals being discussed in this chapter. A high view of Scripture was the reason for the formation of the Evangelical Theological Society in 1949. The authority of Scripture was the main issue in the conservative resurgence in the Southern Baptist Convention, which caused some moderates and progressive Baptists to form the Cooperative Baptist Fellowship and leave the SBC.[43] The integrity of Scripture unites conservative evangelicals with one another and often divides them from other Christians.

A high view of Scripture forms the basis of evangelical ethics. James Gustafson, writing on the uses of Scripture in moral theology, describes the conservative evangelical understanding of Scripture in ethics as "revealed morality." In this view, he writes, "the 'propositional revelation' of the Bible has an authority in matters of both faith and conduct. The words of the Bible are quite literally the word of God, whether they tell us about God and his glory, about man and his rebellion, about the new life that conversion creates, or about the moral conduct that is required of the children of God."[44] His exemplar for this approach to Scripture and ethics is Carl F. H. Henry.

Henry, who is a key early figure in evangelicalism, argues for the necessity of ethics derived from special revelation. In *Christian Personal Ethics*, he contrasts revelational ethics with speculative ethics, arguing the latter allows for moral positions to change as the situation changes while the former is consistent across time.[45] Henry is also one of the key proponents of the idea of the propositional content of Scripture; he

---

42. Wilkinson, *Between God and Green*, 79.
43. Chute, Finn, and Haykin, *The Baptist Story*, 275–91.
44. Gustafson, "The Changing Use of the Bible in Christian Ethics," 135.
45. Henry, *Christian Personal Ethics*, 161–71.

makes his argument for that position at great length in his multi-volume text, *God, Revelation, and Authority*.[46] A fuller treatment of Henry's well-developed doctrine of Scripture is beyond the scope of this project, but it bears noting that the same individual who wrote six volumes to defend the doctrine of revelation also wrote a volume chastising theologically conservative Christians for relative inaction against the social evils of the world.[47] For Henry, the inerrancy of Scripture does not diminish the need for action, but it is the only solid platform from which just action can be taken. In fact, without inerrancy, the ethical mandates of Scripture bear much less authority.

One common evangelical understanding of biblical inerrancy is summarized at the beginning of the Chicago Statement on Biblical Inerrancy, which reads, "Being wholly and verbally God-given, Scripture is without error or fault in all its teaching, no less in what it states about God's acts in creation, about the events of world history, and about its own literary origins under God, than in its witness to God's saving grace in individual lives."[48] The very next paragraph in the summary statement explains the implications of this doctrine on the authority of Scripture: "The authority of Scripture is inescapably impaired if this total divine inerrancy is in any way limited or disregarded, or made relative to a view of truth contrary to the Bible's own."[49] Much has been written defining the term "inerrancy," discussing the role of textual criticism in validating or undermining inerrancy, and debating the centrality of the doctrine of inerrancy to faithful Christianity. "Inerrancy" tends to be an American term, which evangelicals outside of the US tend to avoid because in those contexts it has a more mechanical connotation. Whether the term "inerrancy" is used or not, among theologically conservative evangelicals the understanding of Scripture as the *norma normata* is central.[50]

The basis of the doctrine of inerrancy for evangelicals is the inspiration of Scripture. For many evangelicals, the appropriate view of God's inspiration of Scripture is that of verbal inspiration. As Robert Gnuse argues, "Strict verbal inspirationists have a very high view of the Bible.

46. Henry, *God, Revelation, and Authority*.

47. Henry, *The Uneasy Conscience of Modern Fundamentalism*.

48. International Council on Biblical Inerrancy, "Chicago Statement on Biblical Inerrancy," 290.

49. International Council on Biblical Inerrancy, "Chicago Statement on Biblical Inerrancy," 290.

50. Bird, *Evangelical Theology*, 62–64.

The words of the Bible are the timeless and eternal communication of God to his people through the mediation of prophets and apostles."[51] Gnuse is critical of inerrantists and presents something of a caricature of the position, though his original definition is sound. He argues that inerrancy is untenable, placing evangelical biblical scholars and theologians into unnecessarily difficult positions. He writes, "Time and energy must be spent defending the amazing accuracy of the Bible and maintaining that in theory this accuracy is perfect . . . As a result conservative scholars have little time to truly address the issues in the text."[52] He goes on to summarize his objection to inerrancy: "The Scriptures are silenced by inerrantists."[53] This last observation runs directly counter to the experience of many evangelical scholars, for whom a high view of Scripture is an enabling and encouraging impetus to dwell in the word of God.

Despite the exaggerated nature of Gnuse's criticism, there is a kernel of explanatory power in it. He is right to say evangelical scholarship is limited by the content of Scripture and compelled to always compare the result of academic study to the content of the canon. This is the norming force of the Bible in evangelical ethics. His criticism also helps to explain the relatively limited output of evangelical scholars on the culturally relevant topic of the environment. As a group of ecotheologians note, biblical scholars like Richard Bauckham are constrained to recover an ecologically positive message from the pages of Scripture rather than simply reading contemporary environmental presuppositions back into the text.[54] Such careful exegesis takes a great deal more work than scholarship that argues for a commonly accepted position on environmentalism and then finds supporting texts or deconstructs conflicting texts to support the accepted position.

Bauckham's methodological approach reflects a high view of the accuracy and authority of Scripture.[55] His methodology is predicated on the value of rightly interpreting the text of the Bible as it was written in the original context and then applying it appropriately to the contemporary setting.[56] Bauckham has published two volumes that combine exacting

51. Gnuse, *The Authority of the Bible*, 25.
52. Gnuse, *The Authority of the Bible*, 31.
53. Gnuse, *The Authority of the Bible*, 31.
54. Horrell, Hunt, and Southgate, "Appeals to the Bible in Ecotheology and Environmental Ethics," 219–38.
55. E.g, Bauckham, *Jesus and the Eyewitnesses*.
56. Bauckham explains his hermeneutical method in careful terms in Bauckham,

exegesis with an interest in rightly caring for creation. In *The Bible and Ecology*, Bauckham careful exegetes texts from Genesis to Revelation, arguing that the text of Scripture itself supports a positive view of the environment.[57] In contrast, there is a more theological emphasis in *Living with Other Creatures*, though Bauckham returns frequently to the text of Scripture to illustrate and, more importantly, support his theological argumentation.[58] Such trenchant exegesis and dogged deference to the text of Scripture reflect a high view of Scripture that empowers Bauckham's environmental ethics, while retaining the central role of the Bible as the highest norm, consistent with his Protestant roots. As evidenced by Bauckham's environmental ethics, the force of the Reformation emphasis on *sola scriptura* is strong in theologically conservative evangelical theology.

Creation

Despite debates on the timing and methods of the origins of the world, theologically conservative evangelicals appear to be united in their understanding of the significant theological influence of the doctrine of creation for environmental ethics.[59] While fundamentalists seem to be united in understanding the doctrine of creation to refer to a defense of creationism within thousands of years of the present day, evangelicals tend to be split on the question of the age of the earth and the means by which the present status of nature came into existence.

The manner in which evangelicals deal with the age of the earth varies greatly. Some evangelicals, like R. J. Berry, are adamant that discussing the age of the earth is counterproductive and actually harms the environment.[60] Other evangelicals, like Stephen Bouma-Prediger, simply assume an ancient earth, largely ignore the question of cosmogony, and

---

The Bible in Politics, 3–19.

57. Bauckham, *Bible and Ecology*.

58. Bauckham, *Living with Other Creatures*.

59. There are a wide variety of terms used to describe the value of creation; however, there is a consistent pattern of valuing creation inherently (as defined in the project) by a range of evangelical theologians writing on creation care: Bauckham, *Living with Other Creatures*, 12–13; Bouma-Prediger, *For the Beauty of the Earth*, 171–72; Wright, "The Earth Is the Lord's," 217–24.

60. Berry, "Disputing Evolution Encourages Environmental Neglect," 113–30.

begin their argument about environmental ethics from that point.[61] In one sense, the assumption of any position is helpful, because it prevents the ethical discussion from being derailed by a finally unresolvable debate. Berry rightly argues that the debate about the age of the earth has the potential to distract from ethical discussions, but he fails to support why this necessarily requires everyone to accept his preferred position on the age of the earth. The more appropriate means of discussing this doctrine, which will be applied in this section, is to focus on the value of the created order instead of attempting to hypothesize the dating of creation.

That being said, the conservative evangelical consensus requires God's involvement in the creative process. Even those who accept an old earth with evolutionary processes leading to the present state of nature tend to view God as the mind behind creation and the sustainer of the created order.[62] A consensus on the details of creation may be unattainable in light of presently accepted scientific data because the exact method of the creative act and means of providence may be undiscernible from the text. But that God did in fact create the world is of paramount significance. As Alistair McGrath notes,

> If the world is indeed created, it follows that the beauty, goodness, and wisdom of its creator are reflected, however dimly, in the world around us. All of us have known a sense of delight at the beauty of the natural world. Yet this is but a shadow of the beauty of its creator. We see what is good, and realize that something still better lies beyond it. And what lies beyond it is not an abstract, impersonal, and unknowable force, but a personal God who has created us in order to love and cherish us.[63]

When humans recognize the relationship between God and his creation, it causes them to see its *inherent* value, that is, its value as it is properly related to God and as it points toward God.[64] But in this description of value lies the danger of seeing creation as having *intrinsic* value. McGrath again points toward the risk that "nature can come to be seen as an end in itself, rather than as a beautiful pointer to a still greater beauty."[65] For

---

61. E.g., Bouma-Prediger, *For the Beauty of the Earth*, 32.
62. For example, Newman, "Progressive Creationism," 103–33.
63. McGrath, *The Reenchantment of Nature*, 18.
64. Meye, "Invitation to Wonder," 33–36.
65. McGrath, *The Reenchantment of Nature*, 19.

McGrath, the value in creation is not *instrumental*; creation's value is in its ability to inspire appreciation of the Creator.[66]

Such an appreciation for God's significant role in sustaining creation helps explain why evangelicals have tended to label their environmental ethics *creation care* instead of environmentalism or ecology.[67] Creation care is a more consistently theological statement than other labels. Christian creation care is not less than popular manifestations of environmentalism; it is more than environmentalism because it sees creation as being imbued with value by its ongoing relationship with the Creator. Thus, as Beisner argues, "The universe is *not* a closed system. Its Creator is distinct from it and—we are not deists, after all—constantly interacting with it."[68] It is the interaction of the Creator with his creation that imparts value to the created order.

Despite variation of terminology, the value of the created order in an evangelical theological perspective can best be described as inherent. Many evangelicals, including McGrath, use the term "intrinsic," but there is a different meaning intended than the self-sufficient value that defines the view for some liberal environmentalists and ecotheologians.[69] Other evangelicals, like Seth Bible and Mark Liederbach, use the terms "inherent" and "intrinsic" interchangeably. What they mean, however, is that creation has value "*apart from the instrumental value* it plays in the life and existence of human beings."[70] There is terminological confusion, but unity of meaning.

There is also unity among evangelicals about the source of value for the created order: the value in creation is dependent upon God. Christopher Wright argues, "In the creation narratives, the affirmation of 'It is good' was not made by Adam and Eve but by God himself. So the goodness of creation (which includes its beauty) is theologically and chronologically prior to human observation. It is something that *God* saw and affirmed before humanity was around to see it."[71]

It bears emphasizing that the evangelical perspective does not, generally, support the sort of destructive domination that results from only

---

66. McGrath, *The Reenchantment of Nature*, xii.
67. Wilkinson, *Between God and Green*, 20.
68. Beisner, *Where Garden Meets Wilderness*, 25–26.
69. McGrath, *The Reenchantment of Nature*, 47.
70. Liederbach and Bible, *True North*, 37. Emphasis original.
71. Wright, *The Mission of God*, 398. Emphasis original.

recognizing the instrumental value. Certainly humans are intended to use the world, based on the evangelical understanding of the value of creation, and they are to use it for the purpose of enjoying God.[72] This recognizes the hierarchical dualism that Bradley Green identifies in Augustine's doctrine of creation and highlights the uniqueness of humans as a special part of creation.[73] However, evangelicals also recognize the potential for abuse of the creation as a result of the sin nature passed down through the fall.[74]

For evangelicals, the world is not the way it is supposed to be. Unlike more anthropologically optimistic perspectives, such as the liberal and ecotheological perspectives, theologians in the evangelical stream accept a literal fall of humanity into sin, which impacts all of creation, not just humanity. Genesis 3:17 is taken as a literal cursing of the earth by God on man's behalf. The future restoration from the effects of the curse is anticipated in Romans 8, but for the present, distortions of the created order are not entirely due to human abuse of the environment.[75] There are certainly evils in creation that are a direct result of ongoing human sin, like degradation of air quality in many cities. However, many of the natural evils in the world are due to the curse, which, according to some interpretations in the evangelical tradition, is intended to point humans toward their need for a restored relationship with the Creator.

This view of creation, which sees direct involvement by God in the origination and sustenance of the created order and the noetic effects of the fall, prevents evangelicals from having the utopian hope for restoration based on human action (or inaction) that some Christians seem to hold.[76] Instead, it leads toward a model which some call "dominion,"

---

72. Meye, "Invitation to Wonder," 30–49.

73. Green, *Colin Gunton and the Failure of Augustine*, 132.

74. See Spencer, "The Inherent Value of the Created Order," 1–17.

75. Some translations render Genesis 3:17 to say that the ground was cursed "because of you" (NASB, NIV, The Message, NLT, ESV, CEV, HCSB), while others read "for thy sake'"(KJV, NKJV, ASV, Young's Literal Translation, Darby Translation). The latter seems a more likely rendering because it matches closely with Romans 8. The former rendering catches one aspect of the text, that man's sin was the cause of the ground's cursing, but it fails to get the second sense of the text, that it was for man's benefit that the ground was cursed. Wolters captures the main point of the passage when stating, "All of creation participates in the drama of man's fall and ultimate liberation in Christ . . . This principle is a clear scriptural teaching," Wolters, *Creation Regained*, 56–57.

76. McFague, *The Body of God*, 30–31.

others "stewardship," and others still different terms. This view describes God as an actor within creation, which shapes the evangelical theological perspective on environmental ethics.

One evangelical voice that differs from others on creation is Calvin Beisner. His position is similar to other evangelicals writing on environmental ethics in many regards except for his valuation of creation. While Beisner affirms the original goodness of creation as it was created, he seems to overestimate the need for humans to alter creation to add to its value. He writes,

> All of the earth was "very good" as God created it ([Gen] 1:31), but Eden was specially good, the Garden within Eden was even more specially good, and the trees of life and of the knowledge of good and evil in the midst of the Garden were still more specially good. Adam's dominion mandate involved his transforming, bit by bit, the rest of the earth from glory to glory . . . The need for transformation by human action does not stand against God's declaration in Genesis 1:31 that all that he had made was very good; it only recognizes that what God made good he intends to make better, and he intends to use people to do it.[77]

His position leads to the criticism from David Horrell, an ecotheologian, that Beisner "favors unfettered economic development as the best means by which developing (as well as developed) countries can increase their wealth and improve their environments. And he does not see any need for those in the richest countries like the USA to reduce their levels of consumption."[78] Elsewhere Horrell asserts, "While environmentalists celebrate the beauty of wildernesses and seek to protect them, Beisner points to the biblical texts that depict wilderness as a sign of God's curse."[79] These comments, coming from a hostile critic, could seem like a strawman—and they are overstatements—but there is sufficient cause for concern in Beisner's environmental ethics to warrant further discussion.

In fact, given that on this point Beisner is critiquing other evangelicals seeking to engage the topic of environmental ethics, it makes sense that he stands apart from the other voices discussed in this section. It would seem that Beisner is arguing that creation has some inherent value to glorify God, but that is dampened significantly by the insistence that

---

77. Beisner, *Where Garden Meets Wilderness*, 13.
78. Horrell, *The Bible and the Environment*, 17.
79. Horrell, *The Bible and the Environment*, 34.

humans can make everything better. There is a narrow criticism to be made here. Horrell seems to be saying that Beisner ascribes only instrumental value to the created order. That is not the case, though Beisner's statements do seem to point to two errors that confuse his understanding of creation's value.

Beisner's first error is in giving insufficient consideration to creation as an actor in giving unmediated glory to God. While Beisner would not deny that the heavens declare the glory of the Lord (cf. Ps 19:1), he seems to be arguing that the created order needs human intervention to best glorify God.[80] That is, somehow creation needs humans to undo the curse.[81] Certainly, humans can work to push back the effects of the curse by limiting suffering and cultivating where necessary, but this does not entail the sort of large-scale taming that Beisner seems to argue for.[82] His purpose for this subjugation is decidedly anthropocentric, but this is likely for the positive reason that he sees humans as the main glorifiers of God.[83] Overcoming the resistance of the cursed earth in the process of subduing it is part of the human penalty for original sin.[84] Thus, Beisner argues that humans are participating in redemption of creation by rolling back the effects of the curse.[85] He seems to be conflating the economic value with the ontological value of an object. Ontologically speaking, both a watch and a tree may give glory to God equally. The watch is an artifact of the *imago Dei* expressed in creative fashion through the mind and hands of the craftsman. The tree reveals the wonder of the mind of the Creator who fashioned and sustained it. Both may have equal ontological value, but the economic value of the watch would be far superior because it is natural resources that have been improved by technology. What Beisner misses is that the tree, apart from human intervention, is still capable of reflecting the glory of God, and thus having inherent ontological value, despite the effects of the curse.

The second error involves underestimating the noetic effects of sin. Here, Beisner fails to account for the unwanted, secondary impact that human intervention in the created order often has. In trying to redeem

---

80. Beisner, *Where Garden Meets Wilderness*, 115–22.
81. Beisner, *Where Garden Meets Wilderness*, 120.
82. Beisner, *Where Garden Meets Wilderness*, 125.
83. Beisner, *Where Garden Meets Wilderness*, 128.
84. Beisner, *Where Garden Meets Wilderness*, 19.
85. Beisner, *Where Garden Meets Wilderness*, 24–25.

creation, humans may gain ground on one front and lose ground on another. This points to a questionable view of the redeemed creation, to which Beisner subscribes in these words:

> In my view, the amazing leaps in economic productivity and human material prosperity stemming from the application of the Christian worldview through the legal, political, economic, scientific, and technological advances propelled by medieval and Reformation churchmen and scientists are a foretaste of the restoration of the cursed creation foretold by Paul and entailed by the incarnation, death, and resurrection of Christ.[86]

Many of the technological advances that Beisner includes are indeed positive for the welfare of humans, but he does not seem to allow sufficient room to critique the manner in which these benefits are obtained. It may be that the advances are a foretaste of restoration, but due to the pervasive influence of sin, technological advances in one area may cause devastating pollution in another. Making life better for humans may not be an unmediated good for the redemption of all creation. It may be that the ethical situation is murkier than either Beisner or his critics allow.

Though Beisner is frequently cited as a representative of an evangelical perspective on the environment, on the question of the value of creation Beisner self-consciously differs from other evangelicals. There is a need for further development on this topic, as Beisner himself admits he is unclear how the implications of his understanding should be worked out.[87] In contrast, most theologically conservative evangelicals tend to value creation inherently, which helps to found a theological perspective for environmental ethics that assigns a stewardship role to humans.

Anthropology

Both the doctrines of revelation and creation feed into an evangelical anthropology. According to Scripture humans were created in the image of God (Gen 1:26–27). There is great debate about what exactly that means, but in evangelical theology, it generally includes the call to become co-creators with God. Humans are called to fulfill the so-called cultural mandate to "be fruitful and multiply and fill the earth and subdue it, and have dominion over the fish of the sea and over the birds of the heavens

---

86. Beisner, *Where Garden Meets Wilderness*, 25.
87. Beisner, *Where Garden Meets Wilderness*, 128.

and over every living thing that moves on the earth" (Gen 1:28). This command is a major point of contention even among evangelicals as the term "dominion" is perceived to have negative connotations. This has led some to interpret the word as "stewardship," though that description is still too strong for some.[88]

The preferred term for human responsibility in creation varies among even the conservative evangelicals who significantly agree on the source of value of creation, the authority of Scripture, and the future fate of creation. For example, Beisner is comfortable using the term "dominion" to refer to human responsibility. As it is defined by many environmentalists, "dominion" really means domination and an assumption of the perverse right to treat creation as if it has only instrumental value.[89] However, according to Beisner, "evangelical environmentalists have sought to answer this charge not by repudiating the doctrine of dominion but by developing a more nuanced understanding of it than the common caricature that sees it as justifying wanton destruction of the natural world."[90] Beisner's position defines the right-most end of the range of evangelical understandings, which helps explain his comfort with the term "dominion."[91] Other evangelicals tend to prefer the term "steward" for the appropriate human role in creation.

According to J. Douma, "A better characterization exists that does justice to both the special position of human beings and the significance of other creatures. I am thinking of the term 'steward.' This designation indeed portrays the theocentric vision. God is the owner; human beings are but stewards tasked with caring for the Lord's possessions."[92] In fact, "stewardship" is the most common term used by evangelicals to indicate the appropriate role of humans with respect to non-human creation.[93] There are significant criticisms of the concept of stewardship from inside

---

88. Bauckham discusses the strengths and weaknesses of the terms in some detail in Bauckham, *Bible and Ecology*, 1–37.

89. Undermining the concept of dominion as a biblical metaphor for the human role in creation is the uniting theme of the essays in Van Houtan and Northcott, eds., *Diversity and Dominion*.

90. Beisner, *Where Garden Meets Wilderness*, 14.

91. His critique of other evangelicals who have written on the environment is tied up in both the value of creation after the fall and human responsibility. Beisner, *Where Garden Meets Wilderness*, 11–23.

92. Douma, *Environmental Stewardship*, 37.

93. Bauckham, *Living with Other Creatures*, 58–62; Fowler, *The Greening of Protestant Thought*, 76.

and outside the evangelical camp.[94] However, stewardship appears to be the most commonly accepted metaphor for the human role in creation among evangelicals because it represents an appreciation of the continuity and discontinuity between humans and the rest of creation.[95]

Scripture seems to support this idea of stewardship. In Eden, Adam was given the responsibility to till and dress the garden (Gen 2:15). Creation was very good, but it was not perfect in the sense that it had reached the end state for which God designed it. Humans were created after the divine recognition that "there was no man to work the ground" (Gen 2:5), implying that humans are designed to add something to God's good creation by their efforts, not simply by their existence. This reflects a notable difference between human and non-human creation: only humans appear to add value to creation by their actions.[96]

This perception is emphasized by Beisner, who argues, "It was Adam's task to transform all of the earth (to subdue and rule it) into a Garden while guarding the original Garden lest it lose some of its perfection and become like the unsubdued earth."[97] For Beisner, this was true even before the fall. The case Beisner makes for this is not grounded on direct exegesis, but it makes some sense in the broader context of Scripture. At the same time, Beisner's concern for development is related to his understanding of the value of creation.

It is not necessary, however, to spend a great deal of energy on conjecture about what the human responsibility beyond the boundaries of Eden may have been without the fall because there are few events recorded in the lives of Adam and Eve before the first sin occurred. Shortly after the fall, God cursed the ground for the sake of the primal couple, which fundamentally changed the relationship between human and non-human creation (Gen 3:17–19). When the fall occurred, all of creation was placed under the curse and the design purpose for creation

---

94. Bauckham offers some criticism: Bauckham, *Bible and Ecology*, 1–36. From outside the evangelical camp, see Northcott, *The Environment and Christian Ethics*, 128–31.

95. E.g., Wilkinson, *Earthkeeping in the Nineties*, 307–23.

96. John Frame sketches a framework for environmental ethics in a similar manner; Frame, *The Doctrine of the Christian Life*, 855–56. This is also consistent with the pattern of biblical texts used in other environmental ethics texts. E.g., Dryness, "Stewardship of the Earth in the Old Testament," 50–65.

97. Beisner, *Where Garden Meets Wilderness*, 13.

was somehow frustrated (cf. Rom 8:19–22).[98] However, the text seems to teach that human responsibility to demonstrate stewardship over the earth did not end when Adam sinned. On the contrary, the necessity for human engagement in creation may have only increased due to human sin.[99] The duty to respond to the effects of the curse illustrates the stewardship relationship of humans with the non-human creation.

Most conservative evangelicals see humans as unique within the created order. For example, Christopher Wright sees humans as kings of creation. He writes, "Accordingly, to rule over the rest of creation as king, to act as the image of God the King, is to do biblical justice in relation to the non-human creation."[100] This kingship is worked out in ecological ethics as a self-sacrificial care for creation.[101] Humans are part of the web of creation, but they stand in distinction from the remainder of creation as having been imbued with the image of God. The preferred nomenclature for evangelical environmental ethics captures the significance of humans within the created order. As Katherine Wilkinson notes, "The term 'creation care,' as opposed to 'environmentalism' or 'sustainability,' engenders concern for God's creation but also maintains a hierarchy within it."[102]

In some circles, hierarchy is equated with abuse. However, this hierarchy does not necessarily engender abuse of the created order—though human sin will often lead to that—but entails the responsibility to care for it. There is a balance between meeting human needs through creation and preserving the integrity of God's created order. On this point especially, evangelicals seem to have a consensus, though there is often debate about how that balance is to be determined and enforced. The general agreement and points of difference all illustrate the value of evaluating a thinker's understanding of the human role in the created order for determining his theological perspective for environmental ethics.

---

98. Wolters, *Creation Regained*, 53–57.

99. Frame notes that the command continued, but he adds that "Rational planning of the use of earth's resources becomes very difficult after the fall, when sinful human beings seek greater portions of the earth at the expense of others." Frame, *The Doctrine of the Christian Life*, 745.

100. Wright, *Old Testament Ethics for the People of God*, 124.

101. Wright, *Old Testament Ethics for the People of God*, 125.

102. Wilkinson, *Between God and Green*, 58.

Eschatology

Evangelicals are not unified on the doctrine of eschatology. One may find nearly every position on the timing of the rapture and the millennium.[103] However, the various interpretations of the sequence of coming events are less significant than the veracity of the predictions and the centrality of an anticipated eschaton to the evangelical Christian; on that point there is general agreement. As Russell Moore writes, "In Scripture the eschaton is not simply tacked onto the gospel at the end. It is instead the vision toward which all of Scripture is pointing—and the vision that grounds the hope of the gathered church and the individual believer."[104] Whatever the evangelical's position on the millennium and timing of the rapture, there is consensus that Scripture says something true about what God will do in the future to redeem his creation.[105]

Speaking in more general terms, Katherine Wilkinson notes that many evangelical "climate care leaders see eschatology as a core theological issue."[106] Overall, she argues,

> eschatological views inform the meaning of the physical world, painting the earth as either a "temporary stage" for or integrated with this biblical drama. In other words, views diverge on the basis of disjunction or continuity between the material present and the future—whether the present creation can be discarded or must be restored. These leaders typically embrace a particular "eschatology of renewal," and paired with a sense of beginnings rooted in Genesis, their views on the end of time frame creation care, book-ending the stewardship story in which they cast themselves and other evangelicals as actors.[107]

Wilkinson connects premillennial dispensationalism with anti-environmental attitudes. However, as discussed above, that is not necessarily the case. Still, Wilkinson is correct in noting the strong influence of eschatology for motivating stewardship among evangelicals.

---

103. Moore, "Personal and Cosmic Eschatology," 708–11.
104. Moore, "Personal and Cosmic Eschatology," 671.
105. Fowler, *The Greening of Protestant Thought*, 77.
106. Wilkinson, *Between God and Green*, 58.
107. Wilkinson, *Between God and Green*, 59.

In an evangelical theological perspective, the return of Christ and the inauguration of the eschaton encourage gospel-oriented social ethics.[108] Moore notes the significance of eschatology for the care of creation:

> God will not wipe out the physical universe, but he will redeem it. We cannot therefore share an economic libertarian's purely utilitarian view of the earth and its resources. Nor can we share a radical environmentalist's apocalyptic scenarios of the world's end "if we don't act now" or his view of human beings as a parasitic presence on the planet . . . We do not worship trees, but we know that we cannot be happy when they are cleared away to make room for one more chain discount store.[109]

Moore's comments illustrate this significance of the question of the end purpose of creation for a theological perspective on environmental ethics. Today's ethical duty is significantly influenced by God's coming judgment and participating as stewards in his efforts on earth.[110]

Recognizing the need to steward ecological resources well, John Frame argues that well-reasoned environmental policies are made much more difficult because of the fall, "so the chief need of the environment is evangelism."[111] Thus, for Frame, personal eschatology is part of the solution for the environment, as redeemed people will live more closely to the creational design and better reflect the cohabitation of the eschaton. However, the cosmic nature of the eschatological redemption is a more common theme among evangelicals speaking of the environment.

The connection between cosmic redemption and ecological sensitivity can be witnessed especially in Albert Wolters, *Creation Regained*. The book is often used as an introductory text for environmental ethics courses, when it is, in fact, a book on Reformed worldview. This is because, for Wolters, "the redemption of Jesus Christ means the *restoration* of an original good creation . . . In other words, redemption is

---

108. Frame argues, "Scripture, as I pointed out, has little to say about the millennium and its supposed ethical implications. But it does say much about the ethical implications of the return of Christ and the final judgment. Indeed, Scripture's main use of these doctrines is ethical. It does not teach us about Jesus' return primarily to stimulate us to draw charts or to determine the precise order of events on the last day, but to show us how to live. It is remarkable that almost every text regarding the return of Jesus has an ethical thrust." Frame, *The Doctrine of the Christian Life*, 282.

109. Moore, "Personal and Cosmic Eschatology," 718.

110. Liederbach and Bible, *True North*, 110.

111. Frame, *The Doctrine of the Christian Life*, 745.

recreation."[112] In this worldview, "the whole course of history [is] a movement from a garden to a city."[113] Although Wolters did not set out to write an environmental ethics text, the totality of the vision he presents of a Christian understanding of the world naturally encompasses the purpose of creation. In that sense, though the route has become tortuous due to the fall, eschatology is directly connected to the doctrine of creation because it describes the fulfillment of the divine purposes in creation. Wolters writes,

> Foundational to everything we have been saying [about creation] is the conviction, based on the Bible's testimony, that the Lord does not forsake the work of his hands. In faithfulness he upholds his creation order. Even the great crisis that will come on the world at Christ's return will not annihilate God's creation or our cultural development of it. The new heaven and the new earth the Lord has promised will be a continuation, purified by fire, of the creation we now know.[114]

Wolters speaks from a more Reformed perspective than some evangelicals are comfortable with, but his perspective has been influential for reinforcing the importance of human culture and right stewardship of creation. This vision, while founded in the value of creation, is directed by his eschatology.

Much like Wolters's perspective, Douglas Moo does deep exegetical work to show that the phrase "new creation" in the New Testament refers to the ongoing renewal of the present creation. This renewal is the result of the work of Christ's resurrection and points toward the eventual, dramatic renewal of all things.[115] Though Moo's position differs little from Wolters's, the article is worthy of note because it illustrates implicit responses to the four questions in an evangelical theological perspective for environmental ethics. The meaning of "new creation" is a very minor point, and Moo's take is not novel, but the right interpretation of Scripture is so important that it justifies an entire article to support the meaning of a phrase used just twice in Paul's writings (Gal 6:15, 2 Cor 5:17). While arguing for the eschatological state of the present creation as being renewed, Moo shows that the present creation stands in continuity with

---

112. Wolters, *Creation Regained*, 12. Emphasis original.
113. Wolters, *Creation Regained*, 78.
114. Wolters, *Creation Regained*, 46–47.
115. Moo, "Creation and New Creation," 39–60.

the future one. Therefore, the present creation has value independent of its instrumental worth, which provides appropriate motivation for Christians to be faithful stewards of it.[116] Moo's eschatology of continuity is inseparable from the rest of his theology; thus, a single, brief article has the ability to demonstrate his theological perspective for environmental ethics.

The majority of positive statements on the environment come from evangelicals whose theology is in the Reformed tradition. This is because within the Reformed worldview there tends to be a greater appreciation of the unity between the present creation and the coming new heavens and new earth. However, as Frame notes, the stereotypical assertion that premillenialism necessarily leads to a lack of social concern is countered by the historic examples of "Arminian premils like Jerry Falwell and Pat Robertson."[117] While not known for environmental ethics, there are many examples of non-Reformed premillennialists among fundamentalists who affirm the cultural mandate. Therefore, it is not that non-Reformed eschatologies cannot found ethical concern for the environment; it is simply that there are few, if any, examples in print. This tends to reinforce the assertion that understanding someone's eschatology is vital to understanding their overall theological perspective on the environment. However, an undue emphasis on eschatology in discussions of environmental ethics also has the possibility of further alienating some, like dispensationalist Christians, from the ongoing conversation about creation care, because creation care is improperly framed as inconsistent with a particular theological schema.

CONCLUSION

As with all of the theological streams discussed in this project, there is variegation among evangelicals on these four doctrinal questions that form a theological perspective for environmental ethics. However, the segment of conservative evangelicals discussed in this chapter are consistent advocates for a high view of Scripture and its central authority for every aspect of life. Science is not discarded as insignificant, but it is not the prime source of authority for evangelicals. Although at times evangelicals use the term "intrinsic," creation tends to be valued inherently as

---

116. Douglas J. Moo, "Creation and New Creation," 60.
117. Frame, *The Doctrine of the Christian Life*, 280.

well as instrumentally. Evangelicals generally favor a stewardship model for creation care that assumes human impact on the environment is both necessary and acceptable but also seeks to use resources wisely and handle the created order carefully. Eschatology is a more divided position for evangelicals than the other three questions; however, evangelicals tend to speak in terms of divine renewal as the end state of creation. Whether that renewal occurs through cataclysmic destruction and recreation or dramatic purging and restoration is a point of division. The consensus of evangelicals speaking is that the final restored state of the created order should inspire action toward reasonably approximating that end now through social action. These positions, which generally represent the position of evangelicals, can be seen developed in some detail in the theology of Francis Schaeffer.

CHAPTER NINE

THE ENVIRONMENTALISM OF FRANCIS SCHAEFFER

ALTHOUGH HE WAS NOT an academic, Francis Schaeffer remains one of the most significant evangelical voices on environmental ethics. His volume, *Pollution and the Death of Man*, was the first major response to Lynn White's essay from a theologically conservative Christian perspective. The basic theological framework of the book stands as a solid example of an evangelical theological perspective for environmental ethics. Though many of the illustrations in the volume are outdated, since it was first published in 1970, it remains a classic in the field. Additionally, Schaeffer should be included in this analysis because he wrote a corpus large enough to answer the four central questions of a theological perspective for environmental ethics from multiple sources across a range of time, which tends to deepen the analysis of his understanding as it can be evaluated against the rest of his worldview.

Though Schaeffer only wrote one small volume on environmental ethics, he was a major proponent of engaging culture through every aspect of his life. *Pollution and the Death of Man* was one of his earlier books, as he began his writing career at the end of the 1960s. With the exception of the 2008 postmortem publication of his studies on Romans 1–8, the last of his books was published in the early 1980s before Schaeffer's untimely death due to cancer.

Schaeffer's theological perspective for environmental ethics is clearly evidenced in *Pollution and the Death of Man*. He has chapters directly dealing with the doctrines of creation, anthropology, and eschatology, answering three of the four questions proposed as forming a theological perspective for environmental ethics in this project. This early entry into the environmental debate assumes rather than defends the overarching authority of the Bible in the environmental debate.[1] His broader corpus provides ample evidence of his position on the authority of Scripture which was then no less marginalized than it is today.

REVELATION

The doctrine of revelation is a strength in the work of Francis Schaeffer. He witnessed the inerrancy controversies in the Presbyterian Church and was alive during the battles over modernism. His work demonstrates keen awareness of the importance of inerrancy for the authority of Scripture in the life of the believer. In his doctrine of Scripture, Schaeffer is in line with the conservative streams of Christianity, including the fundamentalists and the evangelicals as defined here. In fact, Schaeffer argues that inerrancy of Scripture is central to the identity of evangelicals. In *No Final Conflict*, he writes, "We must say that if evangelicals are to be evangelicals, we must not compromise our view of Scripture. There is no use in evangelicalism seeming to get larger and larger, if at the same time appreciable parts of evangelicalism are getting soft at that which is the central core—namely, the Scriptures."[2] In his book *The Church at the End of the Twentieth Century*, Schaeffer explains the significance of the inerrancy of Scripture: "Overwhelmingly, the [significance] is that with the Bible being what it is, God's word as absolute, as God's objective truth, we do not need to be, *and we should not be*, caught in the ever-changing fallen cultures which surround us. Those who do not hold the inerrancy of Scripture do not have this high privilege. To some extent, they are at the mercy of the fallen, changing culture."[3]

The authority of Scripture over all of life is central to Schaeffer's environmental ethics, just as it is over his whole theology. Schaeffer does not develop the doctrine of revelation deeply in *Pollution and the Death*

---

1. Schaeffer addresses briefly the inadequacies of science that denies the role of God in creation. Schaeffer, *Complete Works*, "Pollution and the Death of Man," 5:50.

2. Schaeffer, *Complete Works*, "No Final Conflict," 2:121.

3. Schaeffer, *Complete Works*, "The Church at the End of the Twentieth Century," 4:106. Emphasis original.

of Man, but he does demonstrate that the Bible is a primary theological source for his environmental concern. He seeks a "biblical view of nature,"[4] relies on parables to understand the unity of the human race,[5] and points toward an eschatological vision of the restoration of creation that is directly drawn from Scripture.[6]

Yet, Schaeffer's theology does not affirm *only* Scripture as a source. He recognizes revelation from "the internal nature of man himself which speaks of God as personal, and in the evidence of the thought of God as it is expressed in the external, created universe."[7] In fact, Schaeffer goes on to declare, "General revelation and special revelation constitute a unified revelation."[8] This position affirms the significance of both forms of revelation, while recognizing the overarching authority of Scripture. True information could be gained from general revelation, but the Bible is the key to unlocking and interpreting that knowledge.[9] The metaphor of a key and lock is intentional and central to Schaeffer's argument. He believes that truth *can* be obtained from various extra-biblical sources like reason, experience, and tradition. However, Scripture contains what he calls "true truth," which is used to interpret and unlock the truths of general revelation.[10] In Schaeffer's environmental ethics, this sets up a robust role for science that does not conflict with his appreciation for the centrality of the Bible.

In Schaeffer's *Pollution and the Death of Man*, he relies on testimony from those involved in the study of the environment. Therefore, the testimonies of biologist David B. Wingate against DDT's effects on bird populations and of Thor Heyerdahl regarding the amount of trash in the oceans are accepted data points for the need for a better attitude toward the environment.[11] Since then the importance of DDT in controlling malaria and Zika outbreaks has increased in significance. This has resulted in an ongoing questioning over the wisdom of the Carson-inspired DDT ban, and there is still open debate.[12] At the same time, Heyerdahl's

---

4. Schaeffer, *Complete Works*, "Pollution and the Death of Man," 5:27.
5. Schaeffer, *Complete Works*, "Pollution and the Death of Man," 5:30.
6. Schaeffer, *Complete Works*, "Pollution and the Death of Man," 5:39.
7. Schaeffer, *Complete Works*, "True Spirituality," 3:308.
8. Schaeffer, *Complete Works*, "True Spirituality," 3:308.
9. Schaeffer, *Complete Works*, "Escape from Reason," 1:219.
10. Schaeffer, *Complete Works*, "Escape from Reason," 1:218–23.
11. Schaeffer, *Complete Works*, "Pollution and the Death of Man," 5:3–4.
12. The debate about DDT use is still ongoing. One recent article argues for

grand theories of trans-oceanic migrations of populations have largely been discarded.[13] The point is not that the data that Schaeffer accepts is absolutely correct, but that he finds a way to integrate the contemporary understanding of the world with his biblical worldview in a way that retains the centrality of Scripture. These data support his arguments, but the integrity of his perspective is not based on the scientific evidence used to illustrate possible applications of theological principles.

Not only did Schaeffer accept the validity of the contemporary scientific data, he defended science as a discipline in the face of the attack of Lynn White, who argued that science itself was distorted by Christian anthropocentrism.[14] In contrast, Schaeffer agrees with Richard Means's use of science, though he disputes Means's proposed pantheism as a solution for environmental problems.[15] These careful distinctions within Schaeffer's arguments are significant to his overall theological perspective for environmental ethics.

Schaeffer was willing to say "yes" to one piece of an argument and "no" to another; he was ready to accept the data of science but not to accept the conclusions that others drew from it. Hence, in accepting the problem of indiscriminate spraying of DDT, Schaeffer did not necessarily affirm the solution that others drew from the data: namely the prohibition of the use of DDT, including preventing its use in a controlled fashion to mitigate malaria risks in tropical climates.[16] With Schaeffer there is a willingness to accept scientific data, but never in a manner that

---

continuing to use DDT for mosquito control, due to its positive impact on the community: Bouwman, Berg, and Kylin, "DDT and Malaria Prevention," 744–47. See also a history of the malaria control efforts in Peru, which demonstrates the impact proper usage of DDT had in essentially eliminating the disease: Griffing, Gamboa, and Udhayakumar, "The History of 20th Century Malaria Control in Peru," 1–7. For an opposing viewpoint, which argues DDT use is more costly in the long term, see Pedercini, Blanco, and Kopainsky, "Application of the Malaria Management Model," 1–12.

13. Holton, "Heyerdahl's Kon Tiki Theory and the Denial of the Indigenous Past," 163–81. This article is rather late evidence. There was a great deal of pushback on Heyerdahl's anthropological theories even early on, including accusations of racism.

14. Schaeffer, *Complete Works*, "Pollution and the Death of Man," 5:6.

15. Schaeffer, *Complete Works*, "Pollution and the Death of Man," 5:9–19.

16. Note that Schaeffer does not engage this issue specifically. The point is that he viewed scientific findings through the authority of Scripture. A concrete example of balancing contemporary development and a biblical mandate to care for creation can be found in Schaeffer's discussion of strip mining and electrical wires in the Alps. Schaeffer, *Complete Works*, "Pollution and the Death of Man," 5:48–49.

overrides Scripture.[17] In particular, the ethical implications of Scripture must norm any course of action drawn from contemporary sources. The proper understanding of environmental ethics was, for Schaeffer, drawn from Scripture and largely built on a biblical understanding of the value of the created order.

CREATION

The doctrine of creation for Schaeffer is much more than an account of the number of days it took God to make the world.[18] It is certainly more than an effort to trace back the number of years since the world was created. Schaeffer affirms the historicity of the Genesis account of the creation of the world, but he sees the first book of the Bible as primarily explaining the character of God through the story of creation. The reader can understand God's existence, the order of the universe, the character of God, and the personality of God by reading the Genesis account.[19] More significantly for this project, Schaeffer sees that the creation narrative in Genesis points toward the value of creation, which is one of the four questions in a theological perspective for environmental ethics.[20]

Schaeffer picks up the question of the value of creation directly in *Pollution and the Death of Man*. First, Schaeffer laments the danger of a sort of platonic dualism. In this scheme of value, which Schaeffer associates with many evangelicals, "Nature has become merely an academic proof of the existence of the Creator, with little value in itself. Christians of this outlook do not show an interest in nature *itself*. They use it simply as an apologetic weapon rather than thinking or talking about the real value of nature."[21] Instead of seeing creation as merely a tool for pointing to God, he argues, "It is the biblical view of nature that gives nature a

---

17. For example, Schaeffer often allows for an older earth, as the accepted data of science seems to indicate to many. However, he is always careful to affirm the historicity of the primal couple and the truthfulness of the creation narrative, whatever the timing, because those are central elements of the truthfulness of Scripture. Schaeffer, *Complete Works*, "Genesis in Space and Time," 2:21–46.

18. Schaeffer, *Complete Works*, "Genesis in Space and Time," 2:39.

19. Schaeffer, *Complete Works*, "Genesis in Space and Time," 2:39–40.

20. Schaeffer, *Complete Works*, "Genesis in Space and Time," 2:27–38.

21. Schaeffer, *Complete Works*, "Pollution and the Death of Man," 5:27. Emphasis original.

value *in itself*."²² This begins to sound like intrinsic value, in much the same way that those in the liberal category see nature, though Schaeffer is much more careful to distance himself from pantheism.²³

In fact, Schaeffer uses the term "intrinsic" in describing the value of nature. However, it is clear from the context that Schaeffer is using "intrinsic" to distance his position from mere instrumental value, which he ascribes to modernism.²⁴ He argues, "The man who believes things are there only by chance cannot give things a real intrinsic value. But for the Christian, there is an intrinsic value. The value of a thing is not in itself autonomously, but because God made it. It deserves this respect as something which was created by God, as man himself has been created by God."²⁵ Schaeffer's explanation of "intrinsic" resonates with the definition of inherent value that has been outlined in this project.

Despite Schaeffer's consistent use of the term "intrinsic," it becomes clear through much of his work that he sees the value of the created order as depending on its consistency with its intended function, which is a central aspect of inherent value. He celebrates the architect Frank Lloyd Wright, who sought to design buildings that had integrity with the terrain. He writes, "So there is this desire in our own day to treat material honestly. If we are to have something beautiful, a landscaping that is going to stand with strength, we shall have to keep in mind the integrity of the terrain and the integrity of the material used."²⁶ This view of the inherent value of the created order has direct implications for environmental ethics.

Since creation has native value that is dependent on the integrity of the object's relationship with its design purpose, creation gains rights in Schaeffer's calculus. He pleads, "On a walk in the woods, do not strip the moss from [a rock] for no reason, and then leave the moss to lie by the side and die. Even the moss has a right to live. It is equal with man as a creature of God."²⁷ He goes on to provide several examples that

---

22. Schaeffer, *Complete Works*, "Pollution and the Death of Man," 5:27. Emphasis original.

23. In fact, part of the purpose of *Pollution and the Death of Man* is to debunk Means's argument for Christians becoming pantheists to solve the environmental crisis.

24. Schaeffer, *Complete Works*, "Pollution and the Death of Man," 5:34.

25. Schaeffer, *Complete Works*, "Pollution and the Death of Man," 5:34.

26. Schaeffer, *Complete Works*, "Pollution and the Death of Man," 5:34.

27. Schaeffer, *Complete Works*, "Pollution and the Death of Man," 5:44.

point to the ability of humans to use the created order for purposes that glorify God, but to ultimately respect the integrity of the created order. Therefore, a human may kill an animal for food, but should not slaughter wildlife wantonly. Similarly, trees may be cut down for a housing development, but they should be preserved where possible, even if it raises the cost somewhat to bulldoze around the trees.[28] Thus there is value in creation much the same way humans have value before God, but there is still room for a special place for humans in Schaeffer's theological perspective on environmental ethics.

ANTHROPOLOGY

Schaeffer's anthropology reflects an attempt to remain true to the content of an inerrant Bible. It also shows evidence of efforts to value the created order because the non-human creation has a relationship—albeit one that reflects subject-object interconnection—with the triune Creator. This sense of value in all of creation and solidarity of humans with creation elevates creation. The clear connotation in Scripture of the unique relationship between humans and God permits Schaeffer to assert a hierarchy of rights within the created order that simultaneously limits abuse of non-human creation and enables the use of it for human good.

Schaeffer takes the question of the appropriate role of humans in the created order under direct consideration.[29] As he answers Richard Means's call for a Christian pantheism in *Pollution and the Death of Man*, Schaeffer develops his response through anthropology rather than through theology proper. In other words, Schaeffer does not seek to raise the created order to the status of divinity in an effort to energize its preservation. Instead, he seeks to describe the hierarchy and interconnectedness that rises from the pages of Scripture. Establishing humans as connected to but distinct from the non-human creation is necessary because the sort of pantheism that eliminates the biblical distinctions tends to denigrate humanity rather than exalting the non-human creation.[30] He

---

28. Schaeffer, *Complete Works*, "Pollution and the Death of Man," 5:44.

29. See this question directly asked in these terms: Schaeffer, *Complete Works*, "Pollution and the Death of Man," 5:13.

30. Schaeffer provides the specific example of the pantheism in India, which see humans as equally valuable to rats and cows. Thus animals are allowed to consume resources that could have prevented human starvation. Schaeffer, *Complete Works*, "Pollution and the Death of Man," 5:19.

argues, "Without categories, there is eventually no reason to distinguish bad nature from good nature. Pantheism leaves us with the Marquis de Sade's dictum, 'What is, is right' in morals, and man becomes no more than grass."[31] Schaeffer consistently contests the degradation of humanity through his work. In *Whatever Happened to the Human Race?*, he bases his opposition to abortion on the value of the human life.[32] He founds that understanding of the value of humanity in the Bible, rather than in humanistic philosophy, which he argues inevitably leads to dictation of moral principles by a bare majority.[33] Schaeffer's ethics of environment differs from non-Christian versions—even when it seems to agree at points—because he roots his expression of the proper anthropology in Scripture.[34] The greatest difference between Schaeffer's understanding of a biblical anthropology and many non-Christian understandings is the nuanced portrait of simultaneous continuity and discontinuity between humans and the non-human creation.

Schaeffer believes in a continuity of humans with the rest of creation. He writes, "This is the true Christian mentality. It rests upon the reality of creation out of nothing by God. All things were equally created out of nothing. *All things, including man, are equal in their origin*, as far as creation is concerned."[35] This he puts humans into a relationship with God that is similar to that of the God-tree relationship. At the same time, there is a chasm between humans and the rest of creation. Schaeffer argues, "Man is separated, as personal, from nature because he is made in the image of God."[36] There is thus a distinction between the material relationship of humans with the rest of creation and the unique personal relationship between humans and God, which is enabled by God's special creation of humans in his own image. As Schaeffer notes,

---

31. Schaeffer, *Complete Works*, "Pollution and the Death of Man," 5:19.

32. Schaeffer, *Complete Works*, "Whatever Happened to the Human Race?," 5:284.

33. Schaeffer, *Complete Works*, "Whatever Happened to the Human Race?," 5:286–91.

34. In fact, Schaeffer's insistence on the integrity of the created order tends to sound like Aldo Leopold's land ethic or, perhaps, a more theological version of the land ethic that Wendell Berry presents. This is worth noting, though a more robust discussion of this similarity exceeds the boundaries of the present chapter.

35. Schaeffer, *Complete Works*, "Pollution and the Death of Man," 5:28. Emphasis original.

36. Schaeffer, *Complete Works*, "Pollution and the Death of Man," 5:29.

> On a very different level, we are separated from that which is the "lower" form of creation, yet we are united to it. One must not choose; one must say both. I am separated from it because I am made in the image of God; my integration point is upward, not downward; it is not turned back upon creation. Yet at the same time I am united to it because nature and man are both created by God.[37]

This delicately balanced position keeps Schaeffer from falling prey to the rude domination of creation, of which Christianity is sometimes accused. At the same time, it safeguards Schaeffer from being paralyzed by the co-createdness of humans and trees, which could cause some to devalue humanity in favor of preservation of vegetation.[38] This is a simple anthropology with complex implications, which Schaeffer does not fully develop in his work.

Schaeffer's anthropology is inseparable from his understanding of the value of the created order. With Schaeffer, the principle of createdness is essential. By Schaeffer's lights, it is impossible to outline the proper role for humanity in nature without understanding the design of nature, which is only possible through a creationist cosmology.[39] This createdness entails a proper ordering of the world, which removes the autonomous value from the created order and demands each entity within the created order function according to its design. According to Schaeffer, "God treats his creation with integrity: each thing in its own order, each thing the way he made it."[40] This anticipates the position that "God is treating us like man and expects us to choose and act like man. Thus we must *consciously* deal with the integrity of each thing that we touch."[41]

---

37. Schaeffer, *Complete Works*, "Pollution and the Death of Man," 5:31.

38. Schaeffer fleshes this illustration out more fully. Schaeffer, *Complete Works*, "Pollution and the Death of Man," 5:31–32.

39. This becomes clear in Schaeffer's rejection of the seeming quest within modernity to ascribe autonomy to the material world. Schaeffer, *Complete Works*, "Pollution and the Death of Man," 5:32. Also see Schaeffer, *Complete Works*, "Genesis in Space and Time," 2:40. The term "creationist" is here being used to include positions from young-earth creationism to intelligent design and even theistic evolution. Creationism should be understood to be in opposition to the idea that the present natural order occurred as the result of random chance and time. Rooker and Keathley, *40 Questions About Creation & Evolution*, 15–18.

40. Schaeffer, *Complete Works*, "Pollution and the Death of Man," 5:32.

41. Schaeffer, *Complete Works*, "Pollution and the Death of Man," 5:34. Emphasis original.

Such a position requires an active role for humanity in the created order, which Schaeffer finds in Genesis 2:15.[42] He categorizes this relationship as one of dominion, which may be cause for concern among some environmentalists.

Although Schaeffer uses the term "dominion," his understanding of that term more closely reflects the idea of stewardship, which tends to be more highly regarded in contemporary discussions. Schaeffer explicitly rejects the idea that humans have the right to dominate nature and destroy it. He calls for "a fresh understanding of man's 'dominion' over nature."[43] On this, Schaeffer explains that the creation "belongs to God, and we are to exercise our dominion over these things not as though entitled to exploit them, but as things borrowed or held in trust . . . Man's dominion is under God's dominion."[44] The fallen nature of humanity, however, has caused humans to abuse creation. Thus, Schaeffer argues, "Because [humanity] is fallen, he exploits created things as though they were nothing in themselves, and as though he has an autonomous right to them."[45] Thus Schaeffer rejects domination of nature and calls his readers to respect the integrity of nature.

This call to respect the integrity of creation is, of course, complicated by the corruption of the created order due to the fall. However, while humanity awaits the final restoration of the created order, Schaeffer's anthropology gives humans the responsibility and right to use the inherently valuable non-human creation in a manner consistent with its design.

ESCHATOLOGY

Like many other evangelicals, Schaeffer understands the eschatological vision of the new heavens to be a radical restoration of the created order when Christ comes again. Given the biblicism of Schaeffer's theological perspective, it is little surprise that he roots his hope for the coming restoration of all creation in Romans 8. He notes, "What Paul says there is that when our bodies are raised from the dead, at that time nature too will be redeemed. The blood of the Lamb will redeem man and nature

---

42. Schaeffer, *Complete Works*, "Genesis in Space and Time," 2:67–68.
43. Schaeffer, *Complete Works*, "Pollution and the Death of Man," 5:40.
44. Schaeffer, *Complete Works*, "Pollution and the Death of Man," 5:40.
45. Schaeffer, *Complete Works*, "Pollution and the Death of Man," 5:41.

together."⁴⁶ This restoration of all of creation, including humanity, marries well with the nuanced anthropology Schaeffer describes. Humanity is linked with the rest of the creation by its status as being created; thus both human and non-human aspects of creation may anticipate redemption in a similar fashion, regarding the material aspects of their being. Thus cosmic redemption entails renewal of all matter from the effects of the fall.

Though Schaeffer does not address the concern directly, he does indicate in the same context a difference between the salvation of the human soul and the redemption of the material creation. Referring to Romans 6, he notes the biblical requirement for faith to lead to salvation. Schaeffer argues that Christ's death is sufficient to heal the breach between God and creation, which was caused by the fall.⁴⁷ However, that does not necessarily entail the full restoration of the personal human-God relationship which was disrupted by the fall and which requires individual faith and repentance for healing. There are multiple dimensions of separation of humans from other entities, which leaves open the possibility of diversity in degree of healing of those various separations.⁴⁸ Thus, Schaeffer's nuanced anthropology manages to reconcile cosmic redemption with particular salvation.⁴⁹

While for evangelicals cosmic redemption tends to be forward-looking, personal salvation is perceived to be a present reality with future implications. Schaeffer anticipates the future restoration of all things but argues personal salvation has immediate ethical implications for the Christian. He argues, "Christians who believe the Bible are not simply called to say that 'one day' there will be healing, but that by God's grace, upon the basis of the work of Christ, substantial healing can be a reality here and now."⁵⁰ He applies this to the various relationships of humanity, not least of which is the human relationship with the rest of creation. Schaeffer writes, "On the basis of the fact that there is going to be total redemption in the future, not only of man but of all creation, the Christian

---

46. Schaeffer, *Complete Works*, "Pollution and the Death of Man," 5:37.

47. Schaeffer, *Complete Works*, "Pollution and the Death of Man," 5:38.

48. Schaeffer outlines four different separations: (1) Humanity from God; (2) A person from himself; (3) A human from other humans; (4) Humanity from nature. Schaeffer, *Complete Works*, "Genesis in Space and Time," 2:69–70.

49. For Schaeffer's clear rejection of universalism, see Schaeffer, *Complete Works*, "The God Who Is There," 1:112.

50. Schaeffer, *Complete Works*, "Pollution and the Death of Man," 5:39.

who believes the Bible should be the man who—with God's help and in the power of the Holy Spirit—is treating nature now in the direction of the way nature will be then."[51] Eschatology is thus a driver for all of Schaeffer's ethics, but especially his environmental ethics.

At this point it is pertinent to note that Schaeffer is not arguing for repristination. Indeed, he distances himself from the utopian vision of a return to primitive life without technology by recognizing the human role to work the Garden of Eden. Schaeffer calls for "substantial healing," which "conveys the idea of a healing that is not *perfect*, but nevertheless is real and evident."[52] This substantial healing of the rift between humanity and nature will result in a restoration of an aesthetic appreciation of the created order. The recognition of beauty in God's handiwork will help the Christian value the goodness of creation and treat it appropriately in light of that value.[53] In recognizing beauty and order in creation, humans are able to rightly exercise dominion or stewardship without sacrificing human uniqueness.[54] An eschatological vision of the coming redemption of all creation empowers Schaeffer's environmental ethics, ties together the other three topics in his theological perspective for environmental ethics, and leads to a practical environmentalism that can be lived faithfully by a Christian.

CONCLUSION

Like many frontrunners in emerging areas of theological debate, it would be easy to anachronistically critique Schaeffer based on later developments. The environmental crises of his day had a stronger causal link between human action and observable diminution of the integrity of the environment: trash was evident in the ocean and indiscriminate use of DDT was resulting in the migration of that pesticide up the food chain.[55] His *Pollution and the Death of Man* does not anticipate the nebulous and ubiquitous concerns of climate destabilization that form the focus of many environmental debates today. Additionally, Schaeffer's choice of the term "intrinsic" for the value of the created order and his use of the

---

51. Schaeffer, *Complete Works*, "Pollution and the Death of Man," 5:39.
52. Schaeffer, *Complete Works*, "Pollution and the Death of Man," 5:39.
53. Schaeffer, *Complete Works*, "Pollution and the Death of Man," 5:42.
54. Schaeffer, *Complete Works*, "Pollution and the Death of Man," 5:43.
55. Schaeffer, *Complete Works*, "Pollution and the Death of Man," 5:3–4.

word "dominion" for the human role in creation are cause for consternation for some. His use of both of these terms should not be criticized based on the definition offered by others, but on the way that he defines them as he uses them.

Similarly, given the sometimes overwhelming shift toward environmentalism as a central aspect of the faith, it is possible to find fault in Schaeffer's emphasis on environmental ethics as an indication of a drift from the centrality of evangelism. However, such a criticism must be tempered significantly by his assertion that only "a truly biblical Christianity has a real answer to the ecological crisis. It offers a balanced and healthy attitude to nature, arising from the truth of its creation by God. It offers the hope here and now of substantial healing in nature of some of the results of the fall, arising from the truth of redemption in Christ."[56] Thus, for Christians to ignore their responsibility to teach and enact a biblical environmental ethics is to deny the world a vision of the substantial healing that only Christianity can provide. As Schaeffer argues, "The church ought to be a 'pilot plant,' where men can see in our congregations and missions a substantial healing of all the divisions, the alienations which man's rebellion has produced."[57] It may be that only through orthopraxy can orthodoxy be heard.

In fact, when it comes to a defense of orthodoxy, few can rival the efforts of Schaeffer. His five-volume collected works teem with arguments for central biblical truths. Volumes such as *The Church at the End of the Twentieth Century*[58] and *The Church before the Watching World*[59] attest to his concern for the centrality of right doctrine to the future of Christianity. This focus on right thinking arises from the need for the church to hold onto doctrines so that it could continue to exist.[60] Schaeffer holds orthodoxy to be a necessity for evangelism. Thus, in the same coherent body of work, a content-bound orthodoxy with a passion for evangelism exists alongside environmentalism, which tends to debunk some popular concerns that environmentalism is necessarily non-Christian. Instead, Schaeffer addressed the degradation of the environment because it was a particular issue of concern and one which he perceived some Christians

56. Schaeffer, *Complete Works*, "Pollution and the Death of Man," 5:7.
57. Schaeffer, *Complete Works*, "Pollution and the Death of Man," 5:47.
58. Schaeffer, *Complete Works*, "The Church at the End of the Twentieth Century," 4:3–114.
59. Schaeffer, *Complete Works*, "The Church before the Watching World," 4:115–80.
60. Schaeffer, *Complete Works*, "The God Who Is There," 1:47.

responding to incorrectly, because evangelical Christianity had lost its vision of the world that included rightly valuing the environment.[61] This one issue is not segregated from the rest of Schaeffer's worldview, but is integral to it and located at a point of significant contact with the world. Thus Schaeffer addresses environmentalism by developing a holistic, biblical theological perspective for environmental ethics that presumes a certain understanding of revelation and directly develops the doctrines of creation, anthropology, and eschatology.

---

61. Schaeffer provides the example of the ugliness of a Christian school in contrast to the beauty of a pagan establishment across the valley. Schaeffer, *Complete Works*, "Pollution and the Death of Man," 5:24.

Conclusion

IF I HAVE ACCOMPLISHED nothing else in this book, I hope that the reader has come to understand the wide range of views that can lead to a positive environmental ethics that honors the goodness of creation. For those who have come to believe that traditional Christian orthodoxy cannot coexist with a robust environmentalism, I hope I have proved you wrong. For those seeking to understand how people with different theological perspectives can come to their conclusions, I hope this helps.

In the preceding chapters, I have surveyed environmental ethics from four different points on a spectrum of Christian belief. There are, no doubt, hundreds of other individuals or groupings that could have been used as examples. At the end of the day, there is no typology that will ever be perfect. But that was not the point. The thesis of the project is that divergences between Christian environmental ethics are largely explained by responses to four particular theological questions. Those questions can be captured under the headings of the doctrines of revelation, creation, anthropology, and eschatology.

It would be a mistake to absolutize those four doctrinal headings as if that was all someone needed to know about someone's environmental ethics to put them into a nice, tight box. That is the mistake that a lot of worldview analysis makes. Applied incorrectly, typologies and frameworks like the one that I spent hundreds of pages arguing for can be used to cut off more conversations than they enable.

On the other hand, recognizing categories and frameworks like a theological perspective for environmental ethics can be helpful to enable conversation, if used properly. These four doctrines provide points of contact between all types of environmental ethics, and they often are the points of most significant difference between different views. For example, Patricia MacCormack's book *The Ahuman Manifesto* has a distinctly negative view of humanity. In fact, she writes, "We are simply parts of a thing known as earth."[1] While it is tempting to begin at the more controversial elements of her philosophy, which includes a celebration of cannibalism,[2] it is likely more helpful in dialogue with the position to begin with the underlying issue of anthropology. What is the proper human role in creation? That question logically entails the goodness of humanity as a special part of creation, which would tend to undermine MacCormack's assertion that it would be better for humans to cease to exist.[3] Not that I think it particularly likely that I will ever encounter MacCormack or that I would be successful in convincing her if I met her, but by addressing the root doctrines at the heart of her misanthropy, I have a better chance of showing why my perspective is different and better. This approach also avoids continued chatter about particular policy issues and draws the conversation back to deeper questions; it is a reminder that someone can have good motivations and come to entirely different policy recommendations. At the worst, if I approach MacCormack's ecosophy using the four doctrinal questions as an analytical framework, I can see beyond the shocking nature of some of her proposals to the deeper issues that drive those proposals. There is little point in arguing against cannibalism as part of an earth-positive ethics if there is no agreement that humanity is a special part of creation. That is really the issue under consideration, and the debate must begin there.

The Ahuman Manifesto is on the edge of possible positions on the environment at this time, so real debate with the perspective is unlikely to prove fruitful. However, a closer example might be to consider how I engage my real, physical neighbors when they have questions about my pollinator gardens. Why is it worth the effort to debate with the Home Owner's Association to allow wildflowers in a portion of my lawn? Their expectation is that a good suburban garden would have short, bloom-intensive flowers laid out in neat rows. The untamed (but well weeded) patch of somewhat stalky wild flowers in my garden does not fit their expectations. Aside from the beauty of the flowers, the value of creation and the benefits to the local ecosystem of these tiny waystations in a desert of lawn make it worth it. My theological perspective drives my actions. Being able to articulate the theological ideals that motivate my actions is

---

1. MacCormack, *The Ahuman Manifesto*, 5.
2. MacCormack, *The Ahuman Manifesto*, 157–64.
3. MacCormack, *The Ahuman Manifesto*, 5–7.

more convincing than bare pragmatic arguments, or at least it has proved to be so far.

Another significant reason to think about a theological perspective for environmental ethics (or any other issue) is that the act of framing my thoughts theologically forces me to consider what is driving my actions. As Lints comments, "Those who . . . remain largely ignorant of their matrices [or their theological perspectives] will be the group most likely controlled by them."[4] By articulating a theological perspective, we at least gain the opportunity to critique our actions and live more mindfully in light of what we claim is good, true, and beautiful. We will never do so perfectly, but there is greater self-awareness in the act of articulation.

This exercise also gives us the opportunity to ask whether our actual theological perspective for environmental ethics fits within our broader worldview. If someone believes God created everything through the power of his words in six days and nights, but does not value creation as God's handiwork, there is an inconsistency there that needs resolution. Similarly, if someone believes that humans are made in the image of God, but supports euthanasia and abortion as a means of reducing the impact on the environment, they will need to consider how those views can fit together. We can achieve a greater degree of integrity in our world-and-life-view if we work through these challenging questions one at a time. That is an end worth striving for.

---

4. Lints, *The Fabric of Theology*, 14.

Bibliography

Allen, Diogenes, and Eric O. Springsted. *Philosophy for Understanding Theology.* Louisville, KY: Westminster John Knox, 2007.
Allen, Paul. *Theological Method.* New York: T. & T. Clark, 2012.
Armstrong, Karen. *The Battle for God.* New York: Knopf, 2000.
Arnold, Ron. *At the Eye of the Storm: James Watt and the Environmentalists.* Chicago: Regnery Gateway, 1982.
Augustine. *The City of God.* Translated by Marcus Dods. Peabody, MA: Hendrickson, 2013.
———. *The Confessions.* Translated by Maria Boulding. Vintage Spiritual Classics. New York: Vintage Books, 1998.
———. *Miscellany of Eighty-Three Questions.* Edited by Raymond Canning, translated by Boniface Ramsey. The Works of Saint Augustine: A Translation for the 21st Century. Brooklyn, NY: New City Press, 2008.
———. "The Nature of the Good." In *The Manichean Debate,* edited by Boniface Ramsey, translated by Roland Teske, 325–45. The Works of Saint Augustine: A Translation for the 21st Century. Brooklyn, NY: New City Press, 1990.
———. *On Christian Teaching.* Translated by R. P. H. Green. New York: Oxford University Press, 2008.
Bakken, Peter W. "The Ecology of Grace: Ultimacy and Environmental Ethics in Aldo Leopold and Joseph Sittler." PhD diss., The University of Chicago, 1991.
———. "Nature as a Theater of Grace: The Ecological Theology of Joseph Sittler." In *Evocations of Grace: The Writings of Joseph Sittler on Ecology, Theology, and Ethics,* edited by Steven Bouma-Prediger and Peter W. Bakken, 1–19. Grand Rapids, MI: Eerdmans, 2000.
Band, Yehuda Benzion. *Light and Matter: Electromagnetism, Optics, Spectroscopy, and Lasers.* Hoboken, NJ: John Wiley, 2006.
Barr, James. *Fundamentalism.* Philadelphia: Westminster, 1978.
Barton, Stephan C. "New Testament Eschatology and the Ecological Crisis in Theological and Ecclesial Perspective." In *Ecological Hermeneutics: Biblical, Historical and Theological Perspectives,* edited by David Horrell et al., 266–82. New York: T. & T. Clark, 2010.
Bauckham, Richard. *Bible and Ecology: Rediscovering the Community of Creation.* Waco, TX: Baylor University Press, 2010.
———. *The Bible in Politics: How to Read the Bible Politically.* Louisville, KY: Westminster John Knox, 1989.

———. *Jesus and the Eyewitnesses: The Gospels as Eyewitness Testimony*. Grand Rapids, MI: Eerdmans, 2006.
———. "Joining Creation's Praise of God." *Ecotheology: Journal of Religion, Nature & the Environment* 7.1 (2002) 45–59.
———. *Jude, 2 Peter*. Waco, TX: Word Books, 1983.
———. *Living with Other Creatures: Green Exegesis and Theology*. Waco, TX: Baylor University Press, 2011.
Bauder, Kevin T. "Fundamentalism." In *Four Views on the Spectrum of Evangelicalism*, edited by Andrew David Naselli and Collin Hansen, 19–49. Grand Rapids, MI: Zondervan, 2011.
———. "A Fundamentalist Response to 'Confessional Evangelicalism.'" In *Four Views on the Spectrum of Evangelicalism*, edited by Andrew David Naselli and Collin Hansen, 97–103. Grand Rapids, MI: Zondervan, 2011.
Bauman, Whitney A. *Religion and Ecology: Developing a Planetary Ethic*. New York: Columbia University Press, 2014.
Bebbington, D. W. *Evangelicalism in Modern Britain: A History from the 1730s to the 1980s*. Boston: Routledge, 1989.
Beisner, E. Calvin. *Where Garden Meets Wilderness: Evangelical Entry into the Environmental Debate*. Grand Rapids, MI: Eerdmans, 1997.
Berkouwer, G. C. *General Revelation*. Grand Rapids, MI: Eerdmans, 1955.
Berry, R. J. "Disputing Evolution Encourages Environmental Neglect." *Science & Christian Belief* 25.2 (2013) 113–30.
Berry, Wendell. "Caught in the Middle." In *Our Only World*, 73–96. Berkeley, CA: Counterpoint, 2015.
———. "A Forest Conversation." In *Our Only World*, 21–52. Berkeley, CA: Counterpoint, 2015.
———. *Jayber Crow*. Washington, DC: Counterpoint, 2000.
———. "On Being Asked for a 'Narrative for the Future.'" In *Our Only World*, 167–76. Berkeley, CA: Counterpoint, 2015.
Bevans, Stephen B. *Models of Contextual Theology*. Rev. ed. Maryknoll, NY: Orbis, 2002.
Bible Truths for Christian Schools. Greenville, SC: Bob Jones University Press, 1988.
Bird, Michael F. *Evangelical Theology: A Biblical and Systematic Introduction*. Grand Rapids, MI: Zondervan Academic, 2013.
Blanchard, Kathryn D'Arcy, and Kevin J. O'Brien. *An Introduction to Christian Environmentalism: Ecology, Virtue, and Ethics*. Waco, TX: Baylor University Press, 2014.
Bleckmann, Charles A. "Evolution and Creationism in Science: 1880–2000." *BioScience* 56.2 (2006) 151–58.
Boff, Clodovis. *Theology and Praxis: Epistemological Foundations*. Translated by Robert R. Barr. Maryknoll, NY: Orbis, 1987.
Bouma-Prediger, Steven. *Earthkeeping and Character: Exploring a Christian Ecological Virtue Ethic*. Grand Rapids, MI: Baker Academic, 2020.
———. *For the Beauty of the Earth: A Christian Vision for Creation Care*. 2nd ed. Grand Rapids, MI: Baker Academic, 2010.
———. *The Greening of Theology: The Ecological Models of Rosemary Radford Ruether, Joseph Sittler, and Juergen Moltmann*. American Academy of Religion Series. Atlanta: Scholars Press, 1995.

Bouwman, Hindrik, Henk van den Berg, and Henrik Kylin. "DDT and Malaria Prevention: Addressing the Paradox." *Environmental Health Perspectives* 119.6 (2011) 744–47.

Braaten, Carl E. *Eschatology and Ethics: Essays on the Theology and Ethics of the Kingdom of God*. Minneapolis: Augsburg, 1974.

———. "Gospel of Justification Sola Fide." *Dialog* 15.3 (1976) 207–13.

Brauer, Jerald C. "Special Issue Honoring Joseph Sittler." *Journal of Religion* 54.2 (1974) 97–165.

Brumley, Albert. "I'll Fly Away." In *The Baptist Hymnal*, 601. Nashville: LifeWay Worship, 2009.

———. "This World Is Not My Home." In *Songs & Hymns of Revival*, edited by Jack Trieber, 485. Santa Clara, CA: North Valley Publications, 2009.

Brunner, Daniel L., Jennifer L. Butler, and A. J. Swoboda. *Introducing Evangelical Ecotheology: Foundations in Scripture, Theology, History, and Praxis*. Grand Rapids, MI: Baker Academic, 2014.

Bultmann, Rudolf. "Is Exegesis without Presuppositions Possible?" In *The Hermeneutics Reader: Texts of the German Tradition from the Enlightenment to the Present*, edited by Kurt Mueller-Vollmer, 242–48. New York: Continuum, 1985.

Callicott, J. Baird. *In Defense of the Land Ethic: Essays in Environmental Philosophy*. Albany, NY: State University of New York Press, 1989.

Carson, D. A. *Christ and Culture Revisited*. Grand Rapids, MI: Eerdmans, 2008.

———. *The King James Version Debate: A Plea for Realism*. Grand Rapids, MI: Baker, 1979.

Chafer, Lewis Sperry. *Systematic Theology*. 8 vols. Dallas, TX: Dallas Seminary Press, 1947.

Chaffey, Timothy. "The Rise and Fall of Inerrancy in the American Fundamentalist Movement." *Answers in Depth* 7 (2012).

Chartier, Germain. *Introduction to Optics*. New York: Springer, 2005.

Chute, Anthony L., Nathan A. Finn, and Michael A. G. Haykin. *The Baptist Story: From English Sect to Global Movement*. Nashville: B&H Academic, 2015.

Clark, Gordon Haddon. *New Heavens, New Earth: First and Second Peter*. Jefferson, MD: Trinity Foundation, 1993.

Clausen, Christopher. "Left, Right, and Science." *Wilson Quarterly* 36.2 (Spring 2012) 16–21.

Cole, Stewart G. *The History of Fundamentalism*. Hamden, CT: Archon Books, 1963.

Colella, E. Paul. "Human Nature and the Ethics of C. I. Lewis." *Transactions of the Charles S. Peirce Society* 27.3 (Summer 1991) 299.

Collins, Antoinette. "Subdue and Conquer: An Ecological Perspective on Gen 1:28." In *Creation Is Groaning: Biblical and Theological Perspectives*, edited by Mary L. Coloe, 19–32. Collegeville, MN: Liturgical Press, 2013.

Coloe, Mary L. "Preface." In *Creation Is Groaning: Biblical and Theological Perspectives*, edited by Mary L. Coloe, vii–x. Collegeville, MN: Liturgical Press, 2013.

Cone, James H. *A Black Theology of Liberation*. 20th anniversary ed. Maryknoll, NY: Orbis, 1990.

Conradie, Ernst M. *Angling for Interpretation: A First Introduction to Biblical, Theological and Contextual Hermeneutics*. Stellenbosch, South Africa: Sun Press, 2008.

———. "Biblical Hermeneutics of Liberation: Modes of Reading the Bible in the South African Context, by G West, 1991." *Journal of Theology for Southern Africa* 85 (1993) 61–65.

———. "Biblical Interpretation within the Context of Established Bible Study Groups." *Scriptura* 78 (2001) 442–47.

———. "Climate Change and the Church: Some Reflections from the South African Context." *Ecumenical Review* 62.2 (2010) 159–69.

———. "Climate Change as a Multi-Layered Crisis for Humanity." Conference paper, Climate Change as a Crisis for Humanity Conference, Uniting Theological College, Sydney, Australia, September 20, 2011.

———. *Creation and Salvation: A Companion on Recent Theological Movements*. Zürich: Brill, 2012.

———. *Creation and Salvation: Dialogue on Abraham Kuyper's Legacy for Contemporary Ecotheology*. Studies in Reformed Theology 20. Boston: Brill, 2011.

———. "Creation and Salvation: Revisiting Kuyper's Notion of Common Grace." In *Creation and Salvation: Dialogue on Abraham Kuyper's Legacy for Contemporary Ecotheology*, edited by Ernst M. Conradie, 95–136. Leiden: Brill, 2011.

———. "Creation at the Heart of Mission?" *Missionalia* 38.3 (2010) 380–96.

———. *An Ecological Christian Anthropology: At Home on Earth?* Burlington, VT: Ashgate, 2005.

———. "Eschatology in South African Literature from the Struggle Period (1960–1994)." *Journal of Theology for Southern Africa* 107 (2000) 5–22.

———. *Fishing for Jonah: Various Approaches to Biblical Interpretation*. Bellville, South Africa: University of the Western Cape, 1995.

———. "Healing in Soteriological Perspective." *Religion & Theology* 13.1 (2006) 3–22.

———. "The Heuristic Key of 'Sustainable Community': A Few Notes." *Scriptura* 75 (2000) 345–57.

———. "How Can We Recognize God in the Singing River?" *Religion & Theology* 14 (2007) 147–53.

———. "Interpreting the Bible Amidst Ecological Degradation." *Theology* 112.867 (2009) 199–207.

———. "Justice, Peace, and Care for Creation: What Is at Stake? Some South African Perspectives." *International Review of Mission* 99.2 (2010) 203–18.

———. "Mission as Evangelism and as Development? Some Perspectives from the Lord's Prayer." *International Review of Mission* 94.375 (2005) 557–75.

———. "A Preface on Empirical Biblical Hermeneutics." *Scriptura* 78 (2001) 333–39.

———. "On the Theological Extrapolation of Biblical Trajectories." *Scriptura* 90 (2005) 901–08.

———. "Revisiting the Reception of Kuyper in South Africa." In *Creation and Salvation: Dialogue on Abraham Kuyper's Legacy for Contemporary Ecotheology*, edited by Ernst M. Conradie, 15–54. Leiden: Brill, 2011.

———. "The Road Towards an Ecological Biblical and Theological Hermeneutics." *Scriptura* 93 (2006) 305–14.

———. "The Salvation of the Earth from Anthropogenic Destruction: In Search of Appropriate Soteriological Concepts in an Age of Ecological Destruction." *Worldviews: Environment Culture Religion* 14.2/3 (2010) 111–40.

———. "Towards an Agenda for Ecological Theology: An Intercontinental Dialogue." *Ecotheology: Journal of Religion, Nature, & the Environment* 10. 3 (2005) 281–343.

———. "Towards an Ecological Biblical Hermeneutics: A Review Essay on the Earth Bible Project." *Scriptura* 85 (2004) 123–35.
———. "Towards an Ecological Reformulation of the Christian Doctrine of Sin." *Journal of Theology for Southern Africa* 122 (2005) 4–22.
———. "What Are Interpretive Strategies?" *Scriptura* 78 (2001) 429–41.
———. "What Is Theological about Theological Anthropology?" *Nederduitse Gereformeerde Teologiese Tydskrif* 45.3/4 (2004) 558–72.
———. "What on Earth Did God Create? Overtures to an Ecumenical Theology of Creation." *The Ecumenical Review* 66.4 (2015).
———. "What on Earth Is an Ecological Hermeneutics? Some Broad Parameters." In *Ecological Hermeneutics: Biblical, Historical and Theological Perspectives*, edited by David Horrell et al., 295–311. New York: T. & T. Clark, 2010.
Conradie, Ernst M., and Willis Jenkins. "Editors' Introduction: Ecology and Christian Soteriology." *Worldviews: Environment Culture Religion* 14.2/3 (2010) 107–10.
Conradie, Ernst M., and Louis C. Jonker. *Angling for Interpretation: A Guide to Understand the Bible Better*. Bellville, South Africa: University of the Western Cape, 2001.
———. "Determining Relative Adequacy in Biblical Interpretation." *Scriptura* 78 (2001) 448–55.
Cooper, John W. "Dualism and the Biblical View of Human Beings (1)." *Reformed Journal* 32.9 (1982) 13–16.
Cooperman, Alan, Gregory A. Smith, and Stefan S. Cornibert. *US Public Becoming Less Religious*. Pew Research, 2015. https://www.pewforum.org/2015/11/03/u-s-public-becoming-less-religious/.
Cox, Harvey. *How to Read the Bible*. San Francisco, CA: HarperOne, 2015.
Cromie, Thetis. "Feminism and the Grace-Full Thought of Joseph Sittler." *Christian Century* 97.13 (1980) 406–08.
Cronon, William. "The Human Factor in Environmental Change." In *American Environmentalism: Readings in Conservation History*, edited by Roderick Nash, 17–24. New York: McGraw-Hill, 1990.
Curry-Roper, Janel M. "Contemporary Christian Eschatologies and Their Relation to Environmental Stewardship." *Professional Geographer* 42.2 (1990) 157–69.
Danielsen, Sabrina. "Fracturing over Creation Care? Shifting Environmental Beliefs among Evangelicals, 1984–2010." *Journal for the Scientific Study of Religion* 52.1 (2013) 198–215.
Danker, Frederick W. "2 Peter 3:10 and Psalm of Solomon 17:10." *Zeitschrift für die neutestamentliche Wissenschaft und die Kunde der älteren Kirche* 53.1–2 (1962) 82–86.
Daynes, Byron W., and Glen Sussman. *White House Politics and the Environment: Franklin D. Roosevelt to George W. Bush*. College Station, TX: Texas A&M University Press, 2010.
Dayton, Donald W. "Some Doubts about the Usefulness of the Category 'Evangelical.'" In *The Variety of American Evangelicalism*, edited by Donald W. Dayton and Robert K. Johnston, 245–51. Knoxville, TN: University of Tennessee Press, 1991.
Dayton, Donald W., and Douglas M. Strong. *Rediscovering an Evangelical Heritage: A Tradition and Trajectory of Integrating Piety and Justice*. Grand Rapids, MI: Baker, 2014.

De Vos, Peter, and Loren Wilkinson, eds. *Earthkeeping, Christian Stewardship of Natural Resources.* Grand Rapids, MI: Eerdmans, 1980.

Deweese, Garrett J., and J. P. Moreland. *Philosophy Made Slightly Less Difficult: A Beginner's Guide to Life's Big Questions.* Downers Grove, IL: IVP Academic, 2005.

DeWitt, Calvin B. *Earth-Wise: A Biblical Response to Environmental Issues.* Grand Rapids, MI: CRC, 1994.

DeYoung, Kevin, and Greg Gilbert. *What Is the Mission of the Church? Making Sense of Social Justice, Shalom, and the Great Commission.* Wheaton, IL: Crossway, 2011.

Dorrien, Gary J. *The Making of American Liberal Theology: Crisis, Irony, and Postmodernity 1950–2005.* Louisville: Westminster John Knox, 2006.

Douma, Jochem. *Environmental Stewardship.* Edited by Albert H. Oosterhoff. Translated by Nelson D. Kloosterman. Eugene, OR: Wipf and Stock, 2015.

Dryness, William. "Stewardship of the Earth in the Old Testament." In *Tending the Garden: Essays on the Gospel and the Earth*, edited by Wesley Granberg-Michaelson, 50–65. Grand Rapids, MI: Eerdmans, 1987.

Dyer, Keith. "When Is the End Not the End? The Fate of the Earth in Biblical Eschatology (Mark 13)." In *Earth Story in the New Testament*, edited by Norman Habel and Vicky Balabanski, 44–46. Sheffield, UK: Sheffield Academic, 2002.

Dyson, Freeman. "The Scientist as Rebel." In *Nature's Imagination: The Frontiers of Scientific Vision*, edited by John Cornwell, 1–11. New York: Oxford University Press, 1995.

Earth Bible Team. "Guiding Ecojustice Principles." In *Readings from the Perspective of the Earth*, edited by Norman Habel, 38–53. Cleveland: Pilgrim Press, 2000.

Eaton, Heather. "The Revolution of Evolution." *World Views: Environment Culture Religion* 11.1 (Spring 2007) 6–31.

Eckberg, Douglas Lee, and T. Jean Blocker. "Christianity, Environmentalism, and the Theoretical Problem of Fundamentalism." *Journal for the Scientific Study of Religion* 35.4 (1996) 343.

Edwards, Denis. "Creation Seen in Light of Christ: A Theological Sketch." In *Creation Is Groaning: Biblical and Theological Perspectives*, edited by Mary L. Coloe, 1–18. Collegeville, MN: Liturgical Press, 2013.

———. *Jesus the Wisdom of God: An Ecological Theology.* Maryknoll, NY: Orbis, 1995.

Elazar, Daniel J. "Covenant and Community." *Judaism* 49.4 (2000) 387.

Emerson, Matthew. "Does God Own a Death Star? The Destruction of the Cosmos in 2 Peter 3:1–13." *Southwestern Journal of Theology* 77.2 (2015) 281–94.

Erickson, Millard J. "Biblical Theology of Ecology." In *The Earth Is the Lord's: Christians and the Environment*, edited by Richard D. Land and Louis Moore, 36–54. Nashville: Broadman Press, 1992.

"Faculty Biography of Ernst Conradie." *University of the Western Cape.* https://www.uwc.ac.za/study/all-areas-of-study/departments/department-of-religion-and-theology/people.

Fink, Cary, and Becka A. Alper. *Religion and Science: Highly Religious Americans Are Less Likely Than Others to See Conflict between Faith and Science.* Pew Research, 2015. https://www.pewresearch.org/internet/wp-content/uploads/sites/9/2015/10/PI_2015-10-22_religion-and-science_FINAL.pdf.

Finn, Nathan A. "John R. Rice, Bob Jones Jr., and the 'Mechanical Dictation' Controversy: Finalizing the Fracturing of Independent Fundamentalism." *Journal of Baptist Studies* 6 (2014) 60–75.

Fletcher, Joseph F. *Situation Ethics: The New Morality*. Philadelphia: Westminster, 1966.
Fowler, Robert Booth. *The Greening of Protestant Thought*. Chapel Hill, NC: University of North Carolina Press, 1995.
Fox, Matthew. *The Coming of the Cosmic Christ*. San Fransisco: Harper Collins, 1988.
Fox, Warwick. *Toward a Transpersonal Ecology: Developing New Foundations for Environmentalism*. Albany: State University of New York Press, 1995.
Frame, John M. *The Doctrine of the Christian Life*. Phillipsburg, NJ: P & R., 2008.
Frye, Northrop. *Fables of Identity: Studies in Poetic Mythology*. New York: Harcourt, Brace, & World, 1963.
Frykholm, Amy Johnson. *Rapture Culture: Left Behind in Evangelical America*. New York: Oxford University Press, 2004.
Geertz, Armin W. "Theory, Definition, and Typology: Reflections on Generalities and Unrepresentative Realism." *Temenos* 33 (1997) 29–47.
Geisler, Norman L. *Systematic Theology: In One Volume*. Minneapolis: Bethany House, 2011.
George, Robert. "Theoecology-Definition: Christian Creation Stewardship in a Changing World." *Theoecology Journal* 1.1 (December 2012).
Gilbert, Greg. *What Is the Gospel?* Wheaton, IL: Crossway, 2010.
Gillespie, Michael Allen. *The Theological Origins of Modernity*. Chicago: University of Chicago Press, 2009.
Gnuse, Robert Karl. *The Authority of the Bible: Theories of Inspiration, Revelation, and the Canon of Scripture*. New York: Paulist Press, 1985.
Goheen, Michael W., and Craig G. Bartholomew. *Living at the Crossroads: An Introduction to Christian Worldview*. Grand Rapids, MI: Baker Academic, 2008.
Golding, William. *The Inheritors*. New York: Harcourt, Brace, & World, 1962.
Gorringe, Timothy. "The Trinity." In *Systematic Theology and Climate Change: Ecumenical Perspectives*, edited by Michael S. Northcott and Peter M. Scott, 15–32. New York: Routledge, 2014.
Gottlieb, Roger S. *A Greener Faith: Religious Environmentalism and Our Planet's Future*. New York: Oxford University Press, 2006.
Granberg-Michaelson, Wesley. *Ecology and Life*. Waco, TX: Word, 1988.
———. "Introduction: Identification or Mastery?" In *Tending the Garden: Essays on the Gospel and the Earth*, edited by Wesley Granberg-Michaelson, 1–5. Grand Rapids, MI: Eerdmans, 1987.
———. *Redeeming the Creation: The Rio Earth Summit: Challenges for the Churches*. Geneva: WCC Publications, 1992.
Green, Bradley G. *Colin Gunton and the Failure of Augustine: The Theology of Colin Gunton in Light of Augustine*. Eugene, OR: Pickwick, 2011.
Green, Gene L. *Jude and 2 Peter*. Grand Rapids, MI: Baker Academic, 2008.
Gregersen, Niels Henrik. "Christology." In *Systematic Theology and Climate Change: Ecumenical Perspectives*, edited by Michael S. Northcott and Peter M. Scott, 33–50. New York: Routledge, 2014.
Griffing, Sean M., Dionicia Gamboa, and Venkatachalam Udhayakumar. "The History of 20th Century Malaria Control in Peru." *Malaria Journal* 12.1 (2013) 1–7.
Gunton, Colin E. *The Triune Creator: A Historical and Systematic Study*. Edinburgh Studies in Constructive Theology. Grand Rapids, MI: Eerdmans, 1998.
Gushee, David P., and Isaac B. Sharp. *Evangelical Ethics: A Reader*. Louisville, KY: Westminster John Knox, 2015.

Gustafson, James M. "The Changing Use of the Bible in Christian Ethics." In *The Use of Scripture in Moral Theology*, edited by Charles E. Curran and Richard A. McCormick, 133–50. New York: Paulist Press, 1984.

———. "The Place of Scripture in Christian Ethics: A Methodological Study." In *The Use of Scripture in Moral Theology*, edited by Charles E. Curran and Richard A. McCormick, 151–77. New York: Paulist Press, 1984.

———. *A Sense of the Divine: The Natural Environment from a Theocentric Perspective*. Cleveland, OH: Pilgrim Press, 1996.

Gutiérrez, Gustavo. *A Theology of Liberation: History, Politics, and Salvation*. Maryknoll, NY: Orbis, 1988.

Habel, Norman. "The Earth Bible Project." *Ecotheology* 5.7 (1999) 123–24.

———. "Introduction." In *Exploring Ecological Hermeneutics*, edited by Norman C. Habel and Peter L. Trudinger, 1–8. Atlanta: Society of Biblical Literature, 2008.

———. "The Origins and Challenges of an Ecojustice Hermeneutic." In *Relating to the Text: Interdisciplinary and Form-Critical Insights on the Bible*, edited by Timothy Sandoval and Carleen Mandolfo, 141–59. London: T. & T. Clark, 2003.

Haidt, Jonathan. *The Righteous Mind: Why Good People Are Divided by Politics and Religion*. New York: Vintage Books, 2013.

Harnad, Stevan. "To Cognize Is to Categorize: Cognition Is Categorization." In *Handbook of Categorization in Cognitive Science*, edited by Henri Cohen and Claire Lefebvre, 19–43. New York: Elsevier, 2005.

Harris, Harriet A. *Fundamentalism and Evangelicals*. New York: Clarendon, 1998.

Healy, Nicholas M. *Hauerwas: A (Very) Critical Introduction*. Grand Rapids, MI: Eerdmans, 2014.

Heffernan, James. "Why Wilderness?: John Muir's 'Deep Ecology.'" In *John Muir, Life and Work*, edited by Sally M. Miller, 102–16. Albuquerque, NM: University of New Mexico Press, 1993.

Hefner, Philip J. *The Scope of Grace: Essays on Nature and Grace in Honor of Joseph Sittler*. Philadelphia: Fortress, 1964.

———. "Sittler, Joseph A., 1906–1987." *Dialog* 27.2 (1988) 82–83.

Heggen, Bruce Allen. "Dappled Things: Poetry, Ecology, and the Means of Grace in Joseph Sittler's Theology for Earth." *Union Seminary Quarterly Review* 51.1–2 (1997) 29–43.

Henry, Carl F. H. *Christian Personal Ethics*. Grand Rapids, MI: Eerdmans, 1957.

———. *God, Revelation, and Authority*. 6 vols. Wheaton, IL: Crossway, 1999.

———. *The Uneasy Conscience of Modern Fundamentalism*. Grand Rapids, MI: Eerdmans, 1947.

Herhold, Robert M. "Probings by Joseph Sittler: Published in Honor of His 75th Birthday." *Christian Century* 96.30 (1979) 915–17.

Heyerdahl, Thor. *Kon-Tiki: Across the Pacific by Raft*. Translated by F. H. Lyon. New York: Rand McNally, 1950.

Hiebert, Paul G. "Critical Contextualization." *International Bulletin of Missionary Research* 11.3 (1987) 104–12.

Holton, Graham. "Heyerdahl's Kon Tiki Theory and the Denial of the Indigenous Past." *Anthropological Forum* 14.2 (2004) 163–81.

Horrell, David. *The Bible and the Environment: Towards a Critical Ecological Biblical Theology*. London: Equinox, 2010.

———. "Introduction." In *Ecological Hermeneutics: Biblical, Historical and Theological Perspectives*, edited by David Horrell et al., 1–12. New York: T. & T. Clark, 2010.
Horrell, David G., Cherryl Hunt, and Christopher Southgate. "Appeals to the Bible in Ecotheology and Environmental Ethics: A Typology of Hermeneutical Stances." *Studies in Christian Ethics* 21.2 (2008) 219–38.
———, eds. *Greening Paul: Rereading the Apostle in a Time of Ecological Crisis*. Waco, TX: Baylor University Press, 2010.
Huebner, Harry John. *An Introduction to Christian Ethics: History, Movements, People*. Waco, TX: Baylor University Press, 2012.
International Council on Biblical Inerrancy. "Chicago Statement on Biblical Inerrancy." *Journal of the Evangelical Theological Society* 21.4 (1978) 289–96.
Jenkins, Willis. *The Future of Ethics: Sustainability, Social Justice, and Religious Creativity*. Washington, DC: Georgetown University Press, 2013.
———. "North American Environmental Liberation Theologies." In *Creation and Salvation: A Companion on Recent Theological Movements*, edited by E. M. Conradie, 273–78. Berlin: LIT, 2012.
———. "Searching for Salvation as Public Theological Exercise: Directions for Further Research." *Worldviews: Environment Culture Religion* 14.2/3 (2010) 258–65.
Johnston, Lucas F. *Religion and Sustainability: Social Movements and the Politics of the Environment*. Bristol, CT: Equinox, 2013.
Jones, Beth Felker. *Practicing Christian Doctrine: An Introduction to Thinking and Living Theologically*. Grand Rapids, MI: Baker, 2014.
Jones, David W., and Russell S. Woodbridge. *Health, Wealth, & Happiness: Has the Prosperity Gospel Overshadowed the Gospel of Christ?* Grand Rapids, MI: Kregel, 2011.
Juschka, Darlene M. "Cane Toads, Taxonomies, Boundaries, and the Comparative Study of Religion." *Method & Theory in the Study of Religion* 16.1 (2004) 12–23.
Kenneson, Phillip D. "There's No Such Thing as Objective Truth, and It's a Good Thing, Too." In *Christian Apologetics in the Post Modern World*, edited by Timothy R. Phillips and Dennis L. Okholm, 155–70. Downers Grove, IL: IVP, 1995.
Kidd, Thomas S. *Who Is an Evangelical? The History of a Movement in Crisis*. New Haven, CT: Yale University Press, 2019.
Kilner, Adam. "Left Behind: A Novel of the Earth's Last Days." *Touchstone* 30.3 (2012) 52–58.
Klein, Ralph W. "Joseph Sittler Remembered." *Currents in Theology and Mission* 16.1 (1989) 5–28.
Kreeft, Peter. *Back to Virtue: Traditional Moral Wisdom for Modern Moral Confusion*. San Francisco: Ignatius, 1992.
Kuhn, Thomas S. *The Structure of Scientific Revolutions*. 50th anniversary ed. Chicago: University of Chicago Press, 2012.
Kureethadam, Joshtrom Isaac. *Creation in Crisis: Science, Ethics, Theology*. Maryknoll, NY: Orbis, 2014.
Kuyper, Abraham. *Scholarship: Two Convocation Addresses on University Life*. Edited and translated by Harry Van Dyke. Grand Rapids, MI: Christian Library Press, 2014.
Land, Richard D., Louis Moore, and L. Russ Bush. *The Earth Is the Lord's: Christians and the Environment*. Nashville: Broadman Press, 1992.

Langford, Michael J. *The Tradition of Liberal Theology*. Grand Rapids, MI: Eerdmans, 2014.
LeFevre, Alphus. "The Gospel Ship." In *Songs & Hymns of Revival*, edited by Jack Trieber, 521. Santa Clara, CA: North Valley Publications, 2009.
Levering, Matthew. *Engaging the Doctrine of Revelation: The Mediation of the Gospel through Church and Scripture*. Grand Rapids, MI: Baker Academic, 2014.
Lewis, C. S. *The Abolition of Man*. San Francisco, CA: HarperSanFrancisco, 2001.
———. *Letters to Malcolm, Chiefly on Prayer*. San Francisco: HarperOne, 1964.
———. "On the Reading of Old Books." In *God in the Dock: Essays on Theology and Ethics*, edited by Walter Hooper, 200–207. Grand Rapids, MI: Eerdmans, 1995.
Lewis, Clarence Irving. *An Analysis of Knowledge and Valuation*. La Salle, IL: Open Court, 1946.
Lewis, Jack P. "The Days of Creation: An Historical Survey of Interpretation." *Journal of the Evangelical Theological Society* 32.4 (1989) 433–55.
"Liberating Life: A Report to the World Council of Churches." in *Liberating Life: Contemporary Approaches to Ecological Theology*, edited by Charles Birch, William Eakin, and Jay B. McDaniel, 273–90. Maryknoll, NY: Orbis, 1990.
Liederbach, Mark, and Seth Bible. *True North: Christ, the Gospel, and Creation Care*. Nashville: B&H Academic, 2012.
Lindbeck, George A. *The Nature of Doctrine: Religion and Theology in a Postliberal Age*. Louisville, KY: Westminster John Knox, 2009.
Lindsey, Hal, and Carole C. Carlson. *The Late Great Planet Earth*. Grand Rapids, MI: Zondervan, 1970.
Lints, Richard. *The Fabric of Theology: A Prolegomenon to Evangelical Theology*. Grand Rapids, MI: Eerdmans, 1993.
Long, Edward Le Roy. *A Survey of Recent Christian Ethics*. New York: Oxford University Press, 1982.
Lowenthal, David. "Introduction." In *Man and Nature*, edited by David Lowenthal, ix–xxix. Cambridge, MA: Harvard University Press, 1967.
Machen, J. Gresham. *Christianity and Liberalism*. Grand Rapids, MI: Eerdmans, 1923.
Maltby, Paul. "Fundamentalist Dominion, Postmodern Ecology." *Ethics & the Environment* 13.2 (Fall 2008) 119–41.
Marie-Daly, Bernice. *Ecofeminism: Sacred Matter Sacred Mother*. Chambersburg, PA: Anima Books, 1991.
MacCormack, Patricia. *The Ahuman Manifesto*. New York: Bloomsbury, 2020.
Marsden, George M. *Fundamentalism and American Culture: The Shaping of Twentieth-Century Evangelicalism 1870–1925*. New York: Oxford University Press, 1980.
Martin, William J. "Special Revelation as Objective." In *Revelation and the Bible: Contemporary Evangelical Thought*, edited by Carl F. H. Henry, 59–72. Grand Rapids, MI: Baker, 1958.
Marty, Martin. "Foreword." In *Evocations of Grace: The Writings of Joseph Sittler on Ecology, Theology, and Ethics*, edited by Steven Bouma-Prediger and Peter W. Bakken, vii–xi. Grand Rapids, MI: Eerdmans, 2000.
———. "Mentor to Many." *Christian Century* 105.3 (1988) 95.
Marty, Martin, and R. Scott Appleby, eds. *Fundamentalisms and the State: Remaking Polities, Economies, and Militance*. Chicago, IL: University of Chicago Press, 1993.
McCune, Rolland D. "The Self-Identity of Fundamentalism." *Detroit Baptist Seminary Journal* 1 (1996) 9–34.

McDaniel, Donald. "Becoming Good Shepherds: A New Model of Creation Care for Evangelical Christians." PhD diss., Southeastern Baptist Theological Seminary, 2011.

McDonagh, Sean. *To Care for the Earth: A Call to a New Theology*. Santa Fe, NM: Bear & Co., 1987.

McFague, Sallie. *The Body of God: An Ecological Theology*. Minneapolis: Fortress, 1993.

McGrath, Alister E. *The Foundations of Dialogue in Science and Religion*. Malden, MA: Blackwell, 1998.

———. *The Reenchantment of Nature: The Denial of Religion and the Ecological Crisis*. New York: Doubleday, 2003.

McKay, Stan. "An Aboriginal Perspective on the Integrity of Creation." In *Ecotheology: Voices from South and North*, edited by David G. Hallman, 213–17. Geneva: WCC Publications, 1994.

McKim, Robert. "On Comparing Religions in the Anthropocene." *American Journal of Theology & Philosophy* 34.3 (2013) 248–63.

Mercer, Calvin. *Slaves to Faith: A Therapist Looks inside the Fundamentalist Mind*. Westport, CT: Praeger, 2009.

Messer, Neil. "Sin and Salvation." In *Systematic Theology and Climate Change: Ecumenical Perspectives*, edited by Michael S. Northcott and Peter M. Scott, 124–40. New York: Routledge, 2014.

Metzger, Bruce M. *A Textual Commentary on the Greek New Testament: A Companion Volume to the United Bible Societies' Greek New Testament*. Peabody, MA: Hendricksen, 1994.

Meye, Robert P. "Invitation to Wonder: Toward a Theology of Nature." In *Tending the Garden: Essays on the Gospel and the Earth*, edited by Wesley Granberg-Michaelson, 30–49. Grand Rapids, MI: Eerdmans, 1987.

Miller, Jon D., Eugenie C. Scott, and Shinji Okamoto. "Public Acceptance of Evolution." *Science* 313.5788 (2006) 765–66.

Moo, Douglas J. "Creation and New Creation." *Bulletin for Biblical Research* 20.1 (2010) 39–60.

Moo, Jonathan A., and Robert S. White. *Let Creation Rejoice: Biblical Hope and Ecological Crisis*. Downers Grover, IL: IVP, 2014.

Moore, Russell. "Heaven and Nature Sing: How Evangelical Theology Can Inform the Task of Environmental Protection (and Vice Versa)." *Journal of the Evangelical Theological Society* 57.3 (2014) 571–88.

———. *Onward: Engaging the Culture without Losing the Gospel*. Nashville: B&H, 2015.

———. "Personal and Cosmic Eschatology." In *A Theology for the Church*, edited by Daniel L. Akin, 671–722. Nashville: B&H, 2014.

———. "Wendell Berry." http://www.russellmoore.com/tag/wendell-berry/.

———. "Why This Election Makes Me Hate the Word 'Evangelical.'" *Washington Post*, February 29, 2016.

Naess, Arne. "Access to Free Nature." *Trumpeter: Journal of Ecosophy* 21.2 (Summer 2005) 48–50.

Nash, James A. *Loving Nature: Ecological Integrity and Christian Responsibility*. Nashville: Abingdon, 1991.

Nelson, Robert H. "Calvinism without God: American Environmentalism as Implicit Calvinism." *Implicit Religion* 17.3 (2014) 249–73.

Nessan, Craig L. *Orthopraxis or Heresy: The North American Theological Response to Latin American Liberation Theology*. Atlanta: Scholars Press, 1989.

———. *The Vitality of Liberation Theology*. Eugene, OR: Pickwick Publications, 2012.

Neuhaus, Richard John. "Liberation as Program and Promise: On Refusing to Settle for Less." *Currents in Theology and Mission* 2.2 (1975) 90–99.

Newman, Robert C. "Progressive Creationism." In *Three Views on Creation and Evolution*, edited by J. P. Moreland and John Mark Reynolds, 103–33. Grand Rapids, MI: Zondervan, 1999.

Newton, Isaac. *Opticks; or, A Treatise of the Reflections, Refractions, Inflections & Colours of Light*. 4th ed. New York: Dover Publications, 1952.

Niebuhr, H. Richard. *Christ and Culture*. Expanded ed. San Francisco: HarperSanFrancisco, 2001.

Northcott, Michael S. "The Dominion Lie: How Millennial Theology Erodes Creation Care." In *Diversity and Dominion: Dialogues in Ecology, Ethics, and Theology*, edited by Kyle Schuyler Van Houtan and Michael S. Northcott, 89–108. Eugene, OR: Cascade, 2010.

———. *The Environment and Christian Ethics*. Cambridge, UK: Cambridge University Press, 1996.

Northcott, Michael S., and Peter Scott, eds. *Systematic Theology and Climate Change: Ecumenical Perspectives*. New York: Routledge, 2014.

———. "Introduction." In *Systematic Theology and Climate Change: Ecumenical Perspectives*, edited by Michael S. Northcott and Peter Scott, 1–14. New York: Routledge, 2014.

———. "The Flowering of Ecotheology." In *Systematic Theology and Climate Change: Ecumenical Perspectives*, edited by Michael S. Northcott and Peter Scott, 124–63. New York: Routledge, 2014.

O'Brien, Kevin J. *An Ethics of Biodiversity: Christianity, Ecology, and the Variety of Life*. Washington, DC: Georgetown University Press, 2010.

O'Donovan, Oliver. "Usus and Fruitio in Augustine, De Doctrina Christiana I." *Journal of Theological Studies* 33.2 (1982) 361–97.

Olson, Roger E. *The Journey of Modern Theology: From Reconstruction to Deconstruction*. Downers Grove, IL: IVP Academic, 2013.

———. *The Story of Christian Theology: Twenty Centuries of Tradition & Reform*. Downers Grove, IL: IVP Academic, 1999.

Outler, Albert Cook. "The Wesleyan Quadrilateral in John Wesley." *Wesleyan Theological Journal* 20.1 (Spring 1985) 7–18.

Pagels, Elaine H. *Adam, Eve, and the Serpent*. New York: Random House, 1988.

Pedercini, Matteo, Santiago Movilla Blanco, and Birgit Kopainsky. "Application of the Malaria Management Model to the Analysis of Costs and Benefits of DDT Versus Non-DDT Malaria Control." *PLoS ONE* 6.11 (2011) 1–12.

Pentecost, J. Dwight. *Things to Come: A Study in Biblical Eschatology*. Findlay, OH: Dunham, 1958.

Peterson, M., et al. *Philosophy of Religion: Selected Readings*. New York: Oxford University Press, 2009.

Plantinga, Alvin. *Knowledge and Christian Belief*. Grand Rapids, MI: Eerdmans, 2015.

———. *Where the Conflict Really Lies: Science, Religion, and Naturalism*. New York: Oxford University Press, 2011.

Primavesi, Anne. *From Apocalypse to Genesis: Ecology, Feminism, and Christianity.* Minneapolis: Fortress, 1991.
Rasmussen, Larry L. *Earth Community Earth Ethics.* Maryknoll, NY: Orbis, 1996.
———. "Luther and a Gospel of Earth." *Union Seminary Quarterly Review* 51.1–2 (1997) 1–28.
Ronnow-Rasmussen, Toni. "Intrinsic and Extrinsic Value." In *The Oxford Handbook of Value Theory*, edited by Iwao Hirose and Jonas Olson, 29–43. New York: Oxford University Press, 2015.
Rooker, Mark F., and Kenneth Keathley. *40 Questions About Creation & Evolution.* Grand Rapids, MI: Kregel, 2014.
Ruether, Rosemary R. "Ecofeminism." In *Ecofeminism and the Sacred*, edited by Carol J. Adams, 13–23. New York: Continuum, 1993.
———. *Gaia & God: An Ecofeminist Theology of Earth Healing.* San Francisco, CA: HarperSanFrancisco, 1992.
———. *Introducing Redemption in Christian Feminism.* Introductions in Feminist Theology. Sheffield, UK: Sheffield Academic, 1998.
———. "Religious Ecofeminism: Healing the Ecological Crisis." In *The Oxford Handbook of Religion and Ecology*, edited by Roger S. Gottlieb, 362–75. New York: Oxford University Press, 2006.
Ryken, Leland. "Formalist and Archetypal Criticism." In *Contemporary Literary Theory: A Christian Appraisal*, edited by Clarence Walhout and Leland Ryken, 1–23. Grand Rapids, MI: Eerdmans, 1991.
Ryrie, Charles Caldwell. *Basic Theology.* Wheaton, IL: Victor Books, 1986.
———. *Dispensationalism Today.* Chicago: Moody Press, 1965.
Salatin, Joel. *The Marvelous Pigness of Pigs: Respecting and Caring for All God's Creation.* Nashville: Faith Words, 2016.
Sandlin, P. Andrew, and John M. Frame. "Reflections of a Lifetime Theologian: An Extended Interview with John M. Frame." In *Speaking the Truth in Love: The Theology of John M. Frame*, edited by John Hughes, 75–110. Phillipsburg, NJ: P. & R., 2009.
Santmire, H. Paul. "In God's Ecology: A Revisionist Theology of Nature." *Christian Century* 117.35 (2000) 1300–1305.
———. *Nature Reborn: The Ecological and Cosmic Promise of Christian Theology.* Minneapolis: Fortress, 2000.
———. *The Travail of Nature: The Ambiguous Ecological Promise of Christian Theology.* Philadelphia: Fortress, 1985.
Sarkar, Sahotra. *Biodiversity and Environmental Philosophy: An Introduction.* New York: Cambridge University Press, 2005.
Schaefer, Jame. *Theological Foundations for Environmental Ethics: Reconstructing Patristic and Medieval Concepts.* Washington, DC: Georgetown University Press, 2009.
Schaeffer, Francis A. *The Complete Works of Francis A. Schaeffer: A Christian Worldview.* 5 vols. Westchester, IL: Crossway Books, 1985.
Schleiermacher, Friedrich. *The Christian Faith.* Berkeley, CA: Apocryphile, 2011.
Schreiner, Thomas R. *1, 2 Peter, Jude.* Nashville: B&H Academic, 2003.
Scott, Peter. "Humanity." In *Systematic Theology and Climate Change: Ecumenical Perspectives*, edited by Michael S. Northcott and Peter M. Scott, 108–23. New York: Routledge, 2014.

Siemsen, Elaine G. *Embodied Grace: Constructing a North American Theology through the Work of Joseph Sittler*. Lewiston, NY: Edwin Mellen Press, 2003.

Simmons, J. Aaron. "Evangelical Environmentalism: Oxymoron or Opportunity?" *Worldviews: Global Religions, Culture & Ecology* 13.1 (2009) 40–71.

Sittler, Joseph. "Called to Unity." *Currents in Theology and Mission* 16.1 (1989) 5–13.

———. *The Care of the Earth*. Minneapolis: Fortress, 2004.

———. "Christ and the Moral Life." *Theology Today* 26.3 (1969) 342–44.

———. "A Christology of Function." *Lutheran Quarterly* 6.2 (1954) 122–31.

———. *The Doctrine of the Word of God: In the Structure of Lutheran Theology*. Philadelphia: United Lutheran Church in America, 1948.

———. "Dogma and Doxa." In *Worship: Good News in Action*, edited by Mandus A. Egge, 7–23. Minneapolis: Augsburg, 1973.

———. "Ecological Commitment as Theological Responsibility." *Southwestern Journal of Theology* 13.2 (1971) 35–45.

———. *The Ecology of Faith: A New Situation in Preaching*. Philadelphia: Muhlenberg, 1961.

———. *Essays on Nature and Grace*. Philadelphia: Fortress, 1972.

———. "Ethics and the New Testament Style." *Union Seminary Quarterly Review* 13.4 (1958) 29–36.

———. *Evocations of Grace: The Writings of Joseph Sittler on Ecology, Theology, and Ethics*. Edited by Steven Bouma-Prediger and Peter W. Bakken. Grand Rapids, MI: Eerdmans, 2000.

———. *Grace Notes and Other Fragments*. Philadelphia: Fortress, 1981.

———. *Gravity and Grace: Reflections and Provocations*. Minneapolis: Augsburg Fortress, 2004.

———. "In the Light of Our Biblical Tradition." In *What Is the Nature of Man? Images of Man in Our American Culture*, 185–94. Philadelphia: Christian Education Press, 1959.

———. "Judeo-Christian Themes: Conflicts and Complements." In *Judaism and the Christian Seminary Curriculum*, edited by J. Bruce Long, 104–08. Chicago: Loyola University Press, 1966.

———. "The Last Lecture: A Walk around Truth, Eternal Life, Faith." *Religion and Intellectual Life* 4.2 (1987) 59–65.

———. "Nature and Grace in Romans 8." In *Evocations of Grace: The Writings of Joseph Sittler on Ecology, Theology, and Ethics*, edited by Steven Bouma-Prediger and Peter W. Bakken, 207–22. Grand Rapids, MI: Eerdmans, 2000.

———. "Nature and Grace: Reflections on an Old Rubric." *Dialog* 3.4 (1964) 252–56.

———. "An Open Letter by Joseph Sittler." *Currents in Theology and Mission* 11.5 (1984) 269–77.

———. "Scope of Christological Reflection." *Interpretation* 26.3 (1972) 328–37.

———. "Space and Time in American Religious Experience." *Interpretation* 30.1 (1976) 44–51.

———. *The Structure of Christian Ethics*. Baton Rouge: Louisiana State University Press, 1958.

———. "A Theology for Earth." In *Evocations of Grace: The Writings of Joseph Sittler on Ecology, Theology, and Ethics*, edited by Steven Bouma-Prediger and Peter W. Bakken, 20–31. Grand Rapids, MI: Eerdmans, 2000.

Skrimshire, Stefan. "Eschatology." In *Systematic Theology and Climate Change: Ecumenical Perspectives*, edited by Michael S. Northcott and Peter M. Scott, 157–74. New York: Routledge, 2014.

Smith, Andrew. "Secularity and Biblical Literalism: Confronting the Case for Epistemological Diversity." *International Journal for Philosophy of Religion* 71 (2012) 205–19.

Smith, Buster G., and Byron Johnson. "The Liberalization of Young Evangelicals: A Research Note." *Journal for the Scientific Study of Religion* 49.2 (2010) 351–60.

Smith, Daniel R. "Toward a Lutheran Theology of Nature: An Ecological Ethics of the Cross." PhD diss., Graduate Theological Union, 2013.

Smith, James K. A. *The Fall of Interpretation: Philosophical Foundations for a Creational Hermeneutic*. Grand Rapids, MI: Baker Academic, 2012.

Snoeberger, Mark A. "Why a Commitment to Inerrancy Does Not Demand a Strictly 6000-Year-Old Earth: One Young Earther's Plea for Realism." *Detroit Baptist Seminary Journal* 18 (2013) 3–17.

Southern Baptist Convention. "Resolution on the Environment, 1970." https://www.sbc.net/resource-library/resolutions/resolution-on-the-environment/

Southgate, Christopher. *The Groaning of Creation: God, Evolution, and the Problem of Evil*. Louisville: Westminster John Knox Press, 2008.

Spencer, Andrew J. "Andrew Fuller and the Doctrine of Revelation." *Southwestern Journal of Theology* 57.2 (2015) 207–26.

———. "Beyond Christian Environmentalism: Ecotheology as an over-Contextualized Theology." *Themelios* 40.3 (December 2015) 414–28.

———. "The Inherent Value of the Created Order: Toward a Recovery of Augustine for Environmental Ethics." *Theoecology Journal* 3 (2014) 1–17.

Stamps, R. Lucas. "A Chalcedonian Argument Against Cartesian Dualism." *Southern Baptist Journal of Theology* 19.1 (2015) 53–66.

Stiling, Rodney, "Natural Philosophy and Biblical Authority in the Seventeenth Century." In *The Enduring Authority of the Christian Scriptures*, edited by D. A. Carson, 115–36. Grand Rapids: Eerdmans, 2016.

Stoll, Mark. *Inherit the Holy Mountain: Religion and the Rise of American Environmentalism*. New York: Oxford University Press, 2015.

———. *Protestantism, Capitalism, and Nature in America*. Albuquerque, NM: University of New Mexico Press, 1997.

Stowell, Joe. "Foreword." In *Which Bible Translation Should I Use? A Comparison of 4 Major Recent Versions*, edited by Andreas J. Köstenberger and David A. Croteau, vii–x. Nashville: B&H Academic, 2012.

Strandberg, Todd. "Bible Prophesy and Environmentalism." *Rapture Ready*. https://www.raptureready.com/2016/08/08/bible-prophecy-vs-the-environment/

Streib, Heinz, and Ralph W. Hood. "Modeling the Religious Field: Religion, Spirituality, Mysticism, and Related World Views." *Implicit Religion* 16.2 (2013) 137–55.

Thompson, Andrew C. "Outler's Quadrilateral, Moral Psychology, and Theological Reflection in the Wesleyan Tradition." *Wesleyan Theological Journal* 46.1 (Spring 2011) 49–72.

Touraine, Alain. *Critique of Modernity*. Translated by David Macey. Cambridge, MA: Blackwell, 1995.

Towns, Elmer L. *Theology for Today*. Mason, OH: Cengage Learning, 2008.

Van Houtan, Kyle Schuyler, and Michael S. Northcott, eds. *Diversity and Dominion: Dialogues in Ecology, Ethics, and Theology*. Eugene, OR: Cascade, 2010.

Van Wieren, Gretel. *Restored to Earth: Christianity, Environmental Ethics, and Ecological Restoration*. Washington, DC: Georgetown University Press, 2013.

Wallis, Jim. *On God's Side: What Religion Forgets and Politics Hasn't Learned About Serving the Common Good*. Grand Rapids, MI: Brazos, 2013.

Wanliss, James. *Resisting the Green Dragon: Dominion, Not Death*. Burke, VA: Cornwall Alliance for the Stewardship of Creation, 2010.

Watt, James. "The Religious Left's Lies." *Washington Post*, May 20, 2005. http://www.washingtonpost.com/wp-dyn/content/article/2005/05/20/AR2005052001333.html.

Weber, Timothy. "Eschatology." In *Dictionary of Christianity in America*, edited by Daniel G. Reid et al., 397–401. Downers Grove, IL: IVP, 1990.

———. "Fundamentalism." In *Dictionary of Christianity in America*, edited by Daniel G. Reid et al., 461–65. Downers Grove, IL: IVP, 1990.

———. "Premillennialism and the Branches of Evangelicalism." In *The Variety of American Evangelicalism*, edited by Donald W. Dayton and Robert K. Johnston, 5–21. Knoxville, TN: University of Tennessee Press, 1991.

Wells, David. "On Being Evangelical: Some Theological Differences and Similarities." In *Evangelicalism*, edited by Mark A. Noll, D. W. Bebbington, and George A. Rawlyk, 389–410. New York: Oxford University Press, 1994.

Wenham, David. "Being 'Found' on the Last Day: New Light on 2 Peter 3:10 and 2 Corinthians 5:3." *New Testament Studies* 33.3 (1987) 477–79.

White, Lynn Townsend. "The Historical Roots of Our Ecologic Crisis." In *Ecology and Life*, edited by Wesley Granberg-Michaelson, 125–37. Waco, TX: Word Books, 1988.

———. "The Historical Roots of Our Ecologic Crisis." In *Ecology and Religion in History*, 15–31. New York: Harper and Row, 1974.

White, R. E. O. *The Changing Continuity of Christian Ethics*. Exeter, UK: Paternoster, 1979.

Wilkinson, Katharine K. *Between God and Green: How Evangelicals Are Cultivating a Middle Ground on Climate Change*. New York: Oxford University Press, 2012.

Wilkinson, Loren. *Earthkeeping in the Nineties: Stewardship of Creation*. Edited by Loren Wilkinson. Rev. ed. Grand Rapids, MI: Eerdmans, 1991.

Williams, Daniel K. *God's Own Party: The Making of the Christian Right*. New York: Oxford University Press, 2010.

Williams, J. Rodman. *Renewal Theology: Systematic Theology from a Charismatic Perspective*. Grand Rapids, MI: Zondervan, 1996.

Wilson, E. O. *The Creation: An Appeal to Save Life on Earth*. New York: Norton, 2006.

Wirzba, Norman. *From Nature to Creation: A Christian Vision for Understanding and Loving Our World*. Grand Rapids, MI: Baker Academic, 2015.

Wolkomir, Michelle, et al. "Denominational Subcultures of Environmentalism." *Review of Religious Research* 38.4 (1997) 325–43.

Wolters, Albert M. *Creation Regained: Biblical Basics for a Reformational Worldview*. Grand Rapids, MI: Eerdmans, 2005.

———. "Worldview and Textual Criticism in 2 Peter 3:10." *Westminster Theological Journal* 49.2 (1987) 405–13.

Woodrum, E., and T. Hoban. "Theology and Religiosity Effects on Environmentalism." *Review of Religious Research* 35.3 (1994) 193–206.

Wright, Christopher J. H. "The Earth Is the Lord's." In *Keeping God's Earth: The Global Environment in Biblical Perspective*, edited by Noah Toly and Daniel Isaac Block, 216–44. Downers Grove, IL: IVP Academic, 2010.

———. *The Mission of God: Unlocking the Bible's Grand Narrative*. Downers Grove, IL: IVP Academic, 2006.

———. *Old Testament Ethics for the People of God*. Downers Grove, IL: IVP Academic, 2004.

Wuthnow, Robert. *Inventing American Religion: Polls, Surveys, and the Tenuous Quest for a Nation's Faith*. New York: Oxford University Press, 2015.

Zaleha, Bernard Daley, and Andrew Szasz. "Why Conservative Christians Don't Believe in Climate Change." *Bulletin of the Atomic Scientists* 71.5 (2015) 19–30.

Zaspel, Fred G. "B. B. Warfield on Creation and Evolution." *Themelios* 35.2 (2010) 198–211.

INDEX

Angling for Interpretation, 52, 65–72, 155.
anthropology, doctrine of, 18–19, 27, 34–36, 59–61, 75–82, 111–15, 127–30, 135, 147, 160–65, 175–76, 200–203, 215–18
Augustine, 25, 27, 29–33, 112, 197

Bauckham, Richard, 171, 193–94, 201–202
Bauder, Kevin, 141, 143–46, 166
Bebbington Quadrilateral, 46, 177, 182–85
Beisner, Calvin, 113–14, 140, 196, 198–202
Berkouwer, G. C., 190
Berry, R. J., 27–28, 162, 166, 174, 194–95
Bevans, Stephen, 49–50, 52, 54
Boff, Clodovis, 49–50, 54
Bouma-Prediger, Stephen, 63, 94–95, 120, 123, 131, 172, 189, 194–95

Chafer, L. S., 150, 155–56, 158–59, 161, 164, 167–70, 176
Cone, James, 42
Conradie, Ernst, 26, 42, 44, 47–48, 51–53, 57, 60, 63–85, 155–56, 188
creation, doctrine of, 10–11, 17–21, 26–34, 51, 53, 55–59, 73–75, 92, 100, 107–111, 116–17, 125–27, 149–50, 160–63, 194–200, 214–15

Daly, Mary, 42
Dorrien, Gary, 120

Earth Bible Project or Team, 50–53, 55, 57–58, 60–62, 70, 73, 75
ecojustice, 44, 50–51, 57–58, 60–62
Ecological Christian Anthropology, An, 42, 60, 63, 75–83; See also Conradie, Ernst
ecotheology, 10–11, 15, 37, 41–85, 140, 153, 181, 184–85, 193
ecotheological, 10, 14, 41–85, 153, 188, 197
Emerson, Mathew Y., 170
eschatology, doctrine of, 18–19, 36–37, 47, 61–64, 70, 82–85, 92–93, 115–17, 130–33, 147, 154, 166–75, 204–207, 220
evangelical (group), 10–11, 14–15, 34, 36–37, 144–45, 153, 181–222

Fabric of Theology, The, 6–7, 154, 185–86, 225; See also Lints, Richard
Fowler, Robert Booth, 4, 23, 45, 91, 94, 96–101, 103–104, 106, 143, 149–50, 153, 176, 201, 204
Frame, John M., 43, 202–203, 205, 207
fundamentalism, 11, 24, 139–78
fundamentalist, 10–11, 15, 23, 25, 27, 68, 139–85

Gottlieb, Roger, 141–43, 181
Gutiérrez, Gustavo, 42

Habel, Norman, 44, 51–53, 190
Henry, Carl F. H., 145–46, 153, 176–77, 191–92
Horrell, David, 15, 48, 51, 55–56, 61–62, 65, 140, 181, 193, 198–99
Hunt, Cheryl, 15, 48, 140, 181, 193

inherent value, 28–34, 92, 110, 135, 194–200, 213–15
instrumental value, 20, 28, 31–34, 110–11, 114, 162–63, 196–201, 207, 214
intrinsic value, 28–34, 50, 57–60, 73–75, 92, 107–111, 195–96, 213–15, 220–21

Jenkins, Willis, 48–49, 56, 65, 93, 168
Journey of Modern Theology, The, 24–25, 41–42, 44, 47, 55, 90–92, 102–103, 105, 141, 157; See also Olson, Roger

Kuyper, Abraham, 23, 26, 71

Langford, Michael, 94–95, 103–104
Late Great Planet Earth, The, 169; See also Lindsey, Hal
Lewis, C. I., 28–31
Lewis, C. S., 12, 54, 100
liberal (theology), 3, 10, 43, 74, 89–135, 141–42, 153, 157, 177, 183, 187–88
liberalism (theological), 89–135
liberation theology, 10, 41–44, 46–48, 50, 54–55, 59, 63, 65, 153, 184; See also ecotheology; praxis
Liederbach, Mark, 34, 196, 205
Lindsey, Hal, 169
Lints, Richard, 6–7, 154, 185–86, 225

MacCormack, Patricia, 34–35, 223–24
Marsden, George, 144, 154–55, 166–67
McGrath, Alister, 103, 157, 195–96
modern, 18, 24, 90–92, 99–107, 116, 121, 126, 135, 141, 154; See also liberal (theology); liberalism (theological)

modernism, 89–96, 98–100, 107, 111–12, 210, 214; See also modern
Moo, Doug, 206–207

Nessan, Craig, 43, 49–50, 54, 56, 63–64
Neuhaus, Richard, 52, 63
Northcott, Michael, 17, 19, 45, 47, 97–98, 104, 115–17, 147–49, 201–202

O'Donovan, Oliver, 32
Olson, Roger, 24–25, 34, 41–44, 47, 55, 90–93, 95–96, 102–103, 105, 108, 115–16, 121, 141, 157; See also *Journey of Modern Theology, The*; *Story of Christian Theology, The*
orthodox and orthodoxy, 10, 42, 47, 54–56, 81, 109–110, 117, 146, 148, 177, 181–222

Pentecost, Dwight, 104, 152, 169–70
Plantinga, Alvin, 101–102, 141
Pollution and the Death of Man, 34, 186, 209–222; See also Schaeffer, Francis
praxis, 10, 46–50, 52, 54, 56, 83, 85, 94; See also ecotheology; liberation theology

revelation, doctrine of, 18–26, 35, 48–57, 66–72, 92–93, 101–106, 123–25, 151–61, 166–67, 183–84, 188–94, 210–13
Ruether, Rosemary Radford, 42–43, 45, 47, 59, 62–63, 94
Ryrie, Charles, 151, 159–61, 163–67

Santmire, Paul, 90, 100, 112–13, 120, 188
Sarkar, Sahotra, 33, 58, 110
Schaeffer, Francis, 12, 16, 34, 45, 97, 148, 186, 190, 208–222
Schleiermacher, Friedrich, 105–106
science, 12, 18–27, 44–45, 55–56, 66, 71, 73, 90–91, 96–105, 111, 133,

INDEX

146, 151, 153, 156–62, 189, 207, 210–13
Scott, Peter, 17, 19, 45, 97, 104, 111
Sittler, Joseph, 15, 89, 94, 118–35
Southgate, Christopher, 15, 48, 55–56, 140, 181, 193
Stoll, Mark, 148, 175, 187
Story of Christian Theology, The, 34, 90, 92–93; See also Olson, Roger
Swoboda, A. J., 45–47, 184–85

"The Historical Roots of Our Ecological Crisis", 4, 100; See also White, Lynn
theological method, 4–8, 22–25, 41, 45–48, 50–54, 73, 96, 122, 184, 186
Towns, Elmer, 76, 156, 160–61, 163
typology, 9–15

value theory, 28–34; See also inherent value; instrumental value; intrinsic value

Wallis, Jim, 182
Watt, James, 172–75
Wesleyan Quadrilateral, 22–23, 26, 54–55, 90, 102, 105, 151, 157, 159, 188–89
White, Lynn, 4, 18, 49, 92, 98–100, 184, 186, 209, 212; See also "The Historical Roots of Our Ecological Crisis"
Wilkinson, Katharine, 45, 140, 149, 173, 181–82, 191, 196, 203–204
Wilson, E. O., 20–21
Wirzba, Norman, 26, 36, 93, 107–108, 149, 169, 174
Wolters, Albert, 171–72, 190, 197, 203, 205–206
Wuthnow, Robert, 182

www.ingramcontent.com/pod-product-compliance
Lightning Source LLC
Chambersburg PA
CBHW050848230426
43667CB00012B/2199